Volume 7

ELECTROMAGNETIC RADIATION

FIREWORK AND FLARE

THE NEW
how it works

Metal cuttings produced by the action of broach teeth are carried away by the cutting oil, which also lubricates the cutting action.
Photo: Paul Brierley.

THE *NEW*
ILLUSTRATED
Science
and
Invention

ENCYCLOPEDIA

H. S. STUTTMAN INC. PUBLISHERS · WESTPORT, CONNECTICUT 06889

Contents

Volume 7

Published by H. S. STUTTMAN INC.
Westport, Connecticut 06889
© Marshall Cavendish Limited 1987, 1989

Library of Congress Cataloging in Publication Data
Main entry under title:
The New Illustrated Science and Invention Encyclopedia
Includes index.
Summary: An alphabetical encyclopedia covering all aspects of science, the physical world, mechanics and engineering.
1. Science—Dictionaries. 2. Engineering—Dictionaries. [1. Science—Dictionaries 2. Engineering—Dictionaries] I. Clarke, Donald. II. Dartford, Mark. III. Title: How It Works.
Q123.I43 1987   503'.21   85-30973
ISBN 0-87475-450-X

Electromagnetic radiation

Many apparently unrelated phenomena, such as light, radio waves and X rays are different examples of just one type of radiation, *electromagnetic* radiation. They are in fact waves of energy produced when an electric charge is accelerated.

A stationary electric charge – in practice, a charged particle such as an electron – is surrounded by *lines of force* which indicate the direction in which another similar charge would move if it were placed near the original charge. If the charge is moved up and down it is decelerated and then accelerated in the opposite direction at each end of the path. These accelerations cause kinks in the lines of force, which move outward from the charge. A moving charge, however, also generates a magnetic field. (This is the principle of electromagnetism, in which the current of moving electrons produces a controllable magnetic field.) An accelerating charge generates a kinked magnetic field, whose lines of force are perpendicular to the electric field. The speed at which these kinks move outward depends on the surrounding material; in a vacuum it is 186,282 miles (299,792 km) per second.

James Clerk Maxwell was the first to calculate that these electromagnetic disturbances could exist, and would travel at this velocity. He noticed that this is the same as the measured speed of light, and suggested that light is a form of electromagnetic radiation.

Frequency and wavelength

To produce a continuous wave of electromagnetic radiation, the charge must be vibrated up and down continuously. The number of vibrations of the charge in one second is called the *frequency* of the resulting wave, and is measured in cycles per second or *Hertz* (Hz), after the scientist who first produced and detected radio waves. The lowest frequencies of interest are around 150,000 Hz (150 kHz), which are *long wave* radio frequencies. VHF radio uses radiation at about 100,000,000 Hz (100 MHz), but light is at very much higher frequencies (600,000 GHz) and X rays higher still (3,000,000,000 GHz).

Another way of distinguishing types of electromagnetic radiation is by their wavelength, the distance between successive crests of the wave. For any type of wave it must be true that velocity of wave = frequency × wavelength.

It was mentioned above that electromagnetic radiation travels at different speeds in different materials, and so the wavelength must also vary according to the medium the wave is passing through – the frequency is always constant. The *wavelength* of a particular radiation usually means the wavelength it would have in a vacuum. For example, the yellow light emitted by a sodium lamp has a wavelength of 589.3 nanometers in a vacuum. (1 nanometer is one thousand millionth of a meter, abbreviated to nm. Wavelengths can also be given in angstroms; 589.3 nm = 5893 A.) In air it is reduced to 589.1 nm, and in glass it is only 388.6 nm.

The longest radio waves are more than 6 miles (over 10,000 m), and the shortest waves (gamma

Right: The Sun is an inexhaustible source producing electromagnetic radiation of almost all wavelengths. Its peak is in the yellow part of the spectrum, characteristic of a body at about 10,832° F. Some stars are much hotter, and appear blue; others have a surface temperature of 5432° F, cooler than a light bulb, though their size gives off a vast amount of heat.

rays) are at wavelengths less than 0.001 nm, smaller than an atom. (Note that low frequencies correspond to long wavelengths, and high frequencies to short wavelengths.) Atomic clocks rely on the constancy of frequencies emitted by excited atoms of particular elements for their accuracy.

Radio waves

Electromagnetic waves longer than 1 mm are known as radio waves. They are subdivided into groups, such as very high frequency (VHF) and ultra high frequency (UHF), depending on their frequency. The very longest waves usually detectable are called VLF, for very low frequency, with wavelengths longer than 6 miles (10 km) and frequencies lower than 30,000 Hz. At this end of the scale, it becomes rather impractical to detect the signals, which are of very low energy.

Below: The entire electromagnetic spectrum. Different detectors are needed for each part, and the eye can only distinguish between the colors in the visible region. Inset: Two "black body" curves are superimposed, showing the wavelengths emitted by the Sun (the temperature of which is 10,832° F) and a bar of an electric radiator (3632° F).

Radio transmitters work on the principle of rapidly switching on and off an electric current. Whenever current is switched on, as in, for example, some domestic appliance, one pulse of electromagnetic radiation – one kink in the lines of force – is produced. If the current is switched on and off at a high frequency, then electromagnetic radiation of that frequency will be produced. This is how radio transmitters work in principle: electrons are forced to pulsate at the chosen frequency along the transmitting antenna, which should be carefully designed for efficient propagation of the chosen wavelength. One of the most popular designs for both transmitters and receivers is the half-wave dipole. This consists of a conductor whose length is half that of the chosen wavelength. Connection to it is normally made by either injecting current into its midpoint (for example, by winding a coil around it or by breaking it and feeding a signal across the two ends) or by injecting a voltage at one of the ends (the most common technique for automobile antennas).

The electrons which constitute the current are accelerated as the current changes, and they radiate electromagnetic waves whose electric field is parallel to the transmitting antenna. If this antenna is vertical, only a vertical antenna will

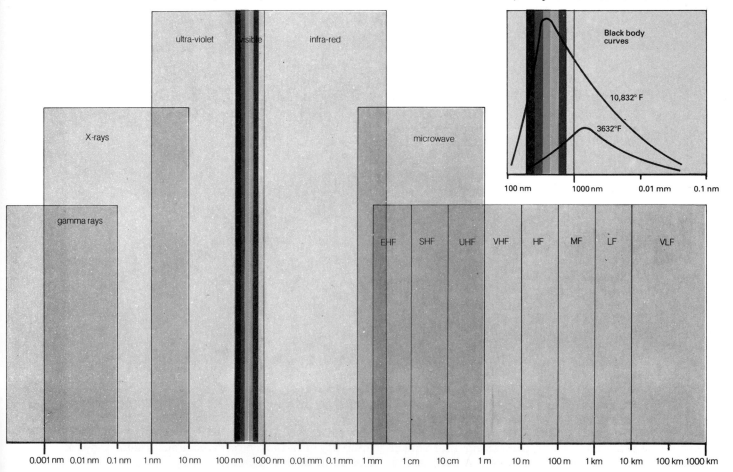

Left: Infrared aerial photograph. Cold areas show up as blue, warmer areas as red. Above: A klystron produces high-frequency radiation which is the basis of TV and microwave transmitters.

receive the radiation; it is said to be vertically *polarized*. Similarly a horizontal antenna will radiate horizontally polarized radiation, which can be detected only by a horizontal receiving antenna. (Polarization can be produced in all forms of electromagnetic radiation, and has useful properties.) The electromagnetic wave produces currents in the receiving antenna which are amplified in the receiver to reproduce the transmitted message.

The highest frequency that can be produced electronically is about 300,000,000,000 Hz, corresponding to a wavelength of 1 mm. Since the term radio usually refers to electronically produced radiation this marks the end of the radio region.

Higher frequencies can be reached by using the natural vibrations of the molecules in a solid. These molecules contain electrons which generate electromagnetic radiation as the molecules vibrate. The hotter the solid is, the more rapidly the molecules vibrate, and the higher the frequency of the radiated electromagnetic waves. Radiation produced in this way is normally unpolarized, because polarization due to electrons moving in different directions will tend to cancel out.

Black-body radiation
Not all the radiation is at the same wavelength, however, even for an object at a constant temperature, and the way in which the energy of the emitted waves changes with frequency is shown in the diagram for bodies at different temperatures. This is a calculated curve, based on the behavior of a theoretical *black body*, which absorbs all the radiation falling on it, and emits it at different wavelengths, but all substances behave something like this if they are placed inside a closed oven. A body as hot

as the Sun radiates most energy at wavelengths around 520 nm, which is the wavelength of yellow-green light, while an object at room temperature emits far less radiation, most of it about 10,000 nm, beyond the visible spectrum, in the infrared. Since the distribution of energy between different wavelengths can be described in terms of the temperature of a black body, the inverse calculation is sometimes done to describe an energy distribution in terms of black-body temperature which, in this case, is referred to as a color temperature.

The infrared lies between visible light and the radio region of the electromagnetic spectrum. The radio wavelengths shorter than 1 m are known as *microwaves*, and they share some properties with infrared. The boundary with the latter occurs at about 1 mm, but the distinction usually made is that radio waves are generated electronically, while infrared is produced thermally.

Infrared is often thought of as heat radiation. The reason for this is not because it has more energy than other wavelengths; the Sun, for example, radiates much more energy as visible light than as infrared. It is because the molecules in objects at room temperature vibrate at about the same frequency as infrared, so the radiation can give up its energy directly to the vibrating molecules. This extra energy makes the molecules vibrate faster, which is felt as heat. Yellow light from the Sun, however, will heat up an object much more than infrared will.

Infrared radiation cannot be detected by the type of receiver used for radio waves, but it can be measured with a BOLOMETER, which measures the total intensity of radiation received, and by some types of electronic detector.

Above: A radiotelescope can receive electromagnetic rays of a very low frequency.

Above: Ultraviolet Sun rays stimulate the pigment melanin which makes the skin tan.

Light

Visible light is at shorter wavelengths than infrared, from 390 to 750 nm. The eye sees different wavelengths as different colors: 680 nm is seen as red; 560 nm, yellow; 500 nm, green; 420 nm, blue; 400 nm, violet. The usual sources of light are hot bodies, such as the Sun or the filament of a tungsten lamp, and light can usually be detected by the human, animal or insect eye, by the camera's photographic plate, or by a photoelectric cell.

To understand why these respond to light but not to infrared radiation, it must be realized that electromagnetic radiation does not travel as a continuous flow of energy, but in bursts of energy, called quanta (or photons). The energy of each quantum depends only on the frequency of the radiation: Energy (in joules) = $66 \times 10^{-38} \times$ frequency (in Hz). For comparison, a 100-watt light bulb emits 100 joules of energy every second. The important thing to notice is that higher frequencies have photons of higher energy.

Both infrared and visible light can be produced by exciting electrons under suitable conditions. Electrons gain discrete quanta of energy from the excitor which they subsequently release as a photon, that is, a packet of light containing just one frequency. Typical environments producing these conditions include sodium lamps, neon signs, lasers, cathode ray tubes and light emitting diodes. Unlike black-body radiators, they can produce light energy without producing large amounts of heat energy.

At the frequency of visible light the energy of a photon is only 4×10^{-19} joules, but this is enough to start some chemical reactions. In the eye the reaction triggers a nerve cell which transmits a message to the brain, whereas in the photographic film emulsion some of the silver compound is changed to silver metal, and the developing process enhances this reaction to produce an image.

Ultraviolet

At wavelengths shorter than 390 nm is the ultraviolet, which extends down to 1 nm. This radiation is emitted by extremely hot bodies, but the temperatures needed are higher than the boiling point of all substances, so ultraviolet is produced this way only in very hot stars.

On Earth ultraviolet is produced in a different way. The electrons in atoms and molecules can have only certain energies, and when they move from one energy state to another they emit the excess energy as electromagnetic radiation. This radiation will occur at particular frequencies corresponding to the energy changes in the atom. Many atoms will produce frequencies which are in the ultraviolet part of the spectrum, one example being mercury, which is used in suntan ultraviolet lamps. Atoms can also produce wavelengths which lie in the visible spectrum by this process. The color of sodium street lights is due to an energy change in the sodium atom which results in radiation whose wavelength corresponds to yellow light.

X rays

Higher frequencies still can be produced by suddenly decelerating a stream of electrons. In a typical apparatus the electrons are suddenly stopped by hitting the metal anode. The wavelength of the radiation emitted can range from 10 nm down to 0.001 nm, depending on how fast the electrons are traveling. These waves are known as X rays.

X radiation is easily detected by a photographic plate: a hospital X-ray examination is recorded on ordinary photographic emulsion. If a picture is not needed, a particle detector can be used. These come in various forms (one being the geiger counter) but the principle is the same for all.

Gamma rays

Shorter wavelengths than these can be reached only by relying on the natural random processes in the nucleus of an atom. A proton in the nucleus can change its energy state, just as the electron in the atom does, but the frequencies emitted will be much higher, corresponding to wavelengths less than 0.01 nm. These waves are called gamma rays, and are the shortest electromagnetic waves yet detected.

Electromagnetic interference

If a source of electromagnetic radiation generates a sufficiently strong electric or magnetic field, it may cause unwanted side-effects in other items of electronic equipment. This phenomenon is called electromagnetic interference (EMI). The commutators of some small electric motors are a common source of EMI and may be responsible for effects such as white spots on a television picture while a vacuum cleaner is in use or crackling from a car radio while the windshield wash pump is running. EMI can be produced by either the magnetic or the electric component of electromagnetic radiation. If a flat conductor is placed across a varying electric field, it will act as one plate of a capacitor and have a voltage induced in it which varies in sympathy with the external electric field. If a linear conductor is placed around a varying magnetic field, it will act like the secondary winding of a transformer and have a current induced in it which will vary in sympathy with the external magnetic field. A common example of this is the hum which can result from hi-fi equipment if the ground wiring is not satisfactory. Adding more ground wires may increase the number of loops created and hence make the problem worse. There are two main techniques for reducing EMI; screening and suppression. Screening attempts to stop radiation from traveling between the source and the susceptible equipment by placing a grounded barrier between them. Depending on the frequencies involved, this will be either a solid metal plate or a wire mesh. Suppression attempts to prevent EMI by using capacitors and inductors to filter out the offending frequencies. Ideally suppression and screening should take place at the source.

The most thorough form of screening involves placing the item to be screened within a sealed metal enclosure called a Faraday cage. Some sensitive measuring equipment needs to be used in Faraday cages with dimensions of several meters, in which case the metal screening material is normally hidden within the walls, floor and ceiling so that the only indication that it is not a conventional room is the complex door sealing arrangements.

Electromagnetic pulse

A massive meteorite impact or a powerful A-bomb detonation will produce a large uncontrolled burst of electromagnetic radiation called electromagnetic pulse (EMP). This energy is distributed throughout most of the electromagnetic spectrum. Initially it travels radially from the source at just under the speed of light (because it is not in a vacuum). Individual wavebands will be reflected, defracted or absorbed by materials which they encounter so that the spectral distribution of energy will vary from place to place.

At distances of more than a few miles from the source, the amounts of light and heat energy received may be insignificant while large amounts of energy are present in other parts of the electromagnetic spectrum. These can produce an extremely strong form of EMI which not only disrupts the operation of electronic circuits but may permanently destroy them.

See also: Electricity; Light; Magnetism; X ray.

Below: A charged particle (1 and 2) has straight or kinked electric lines of force. A bunched wave is produced in magnetic lines of force (3). A vibrating charge (4) creates continuous waves in both electric and magnetic fields.

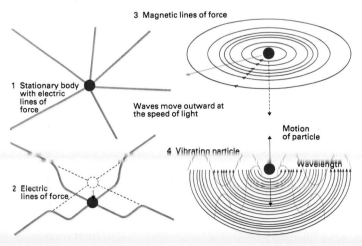

3 Magnetic lines of force

1 Stationary body with electric lines of force

Waves move outward at the speed of light

Motion of particle

4 Vibrating particle

Wavelength

2 Electric lines of force

The beam machine

For many years, scientists have known that light and atomic particles could, separately, provide weapons of unimagined capacity to knock out tanks, aircraft and ships. In the search for new and exciting forms of energy, physicists have shown how near these *beam* weapons are to reality.

Many years of research to harness the atom's power for peace and war have resulted in a vast amount of knowledge about the probable design of a laser gun or particle-beam device, and by the early 1980s, many military scientists were committed to the idea that lasers and particle beam weapons are the only really effective means of combating a nuclear attack.

The basis of a weapon that could prove highly effective against satellites and missiles is microwave radiation – frequently suggested as a means of transmitting energy from orbiting solar power stations to receptors on the ground. Such a weapon would not be intended to burn or blow up targets, because a microwave beam diverges, or spreads out, rapidly and loses power. Instead, beam powers of up to 645 watts per square inch (100 watts per square centimeter) would be used to overload electronic equipment, such as sensors and control circuits, and to jam radar communications.

In the form of *masers*, microwaves can be made powerful, but this technology has largely been superseded by lasers. The word maser is an acronym of Microwave Amplification by Stimulated Emission of Radiation, and masers are the microwave equivalent of lasers, which work by light amplification. The tendency of microwaves to diverge is due to their relatively long wavelength. So the technique of amplifying electromagnetic beams has progressed through the spectrum from masers to infrared lasers and now ultraviolet lasers, both of progressively shorter wavelength.

The potential lethality of lasers comes from the fact that electromagnetic radiation can be amplified and discharged in short, focused pulses, each pulse having a power of many millions of watts. Perfecting a weapons system of this sort is a formidable task, but even more challenging is the problem of aiming the beam. In the battlefield of Earth orbit, for example, the laser light is discharged virtually instantaneously, so there is no scope for mid-course corrections as there is with missiles. Distant targets would need to be selected either electronically or by the aid of high-magnification telescopes, which require extreme stability.

Above: Testing components of a particle-beam weapon. When perfected, it will be able to destroy weapons and satellites in space.

When all these technical problems have been solved, there remains the nuisance that such advanced technology can be foiled by measures such as reflectors and ceramic heat dissipators similar to the tiles that shield the Shuttle from the heat generated during re-entry into the atmosphere.

The next stage in laser sophistication is likely to provide a beam that is not only much more powerful, but also highly penetrating, less prone to divergence and immune to countermeasures. This is the *X raser* – the X-ray equivalent of the laser.

The breakthrough in X-ray lasers came from research into nuclear fusion. Using a technique called *inertial confirmement*, scientists discovered that certain materials can be made to emit beams of coherent, amplified X rays. The technique is to

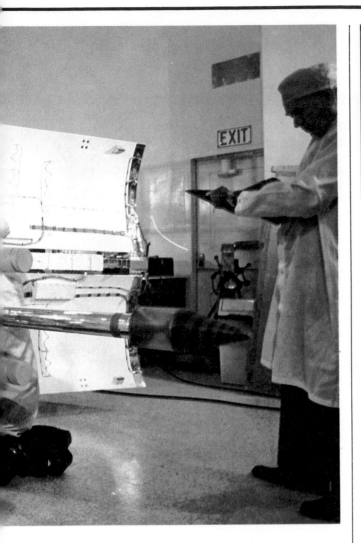

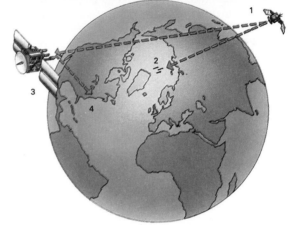

Above: An early-warning satellite (1) detecting an incoming missile (2) alerts headquarters (4) via a satellite link (3). From the control room, a weapon will be directed to destroy the attacking missile.

bombard a sample of carbon simultaneously from several directions, using multiple-beam lasers. The sample becomes heated to the point where all but one of the outer electrons are stripped from the atoms, producing carbon-6.

For laser action, carbon-6 is excited — pumped with energy so that the sample contains more electrons at high energy levels than at the ground, or normal, state. This condition is called *population inversion*. The excited electrons then return to the ground state, emitting photons as they do, but they descend in stages. At one stage, carbon-6 emits a photon of soft X ray, just as an excited ruby atom emits a photon of red laser light. X-ray lasers are less efficient than visible light lasers — they require pumping to extremely high energy levels — but they could generate pulses capable of vaporizing targets at 1243 miles (2000 km).

The most effective antisatellite (Asat) weapon would be one able to fire effectively at all satellites wherever they are located. In that regard, the laser weapon is most efficient. In space, where there is no atmosphere, laser light can be directed with pinpoint accuracy to a target. In the atmosphere, however, the effective range of the weapon is restricted severely. So, a laser-Asat in space can make the best use of the available technology and operate at maximum effect.

Both the U.S.S.R. and the U.S. have been researching laser weapons for many years. Long before public announcement of the intention to proceed with these weapons, both sides had made preparations for a major new role for the military, using space as a theater of war.

Development times are hard to predict for technology that still needs refinement. It is generally agreed by defense scientists that the first laser-Asat weapons could be in orbit by the end of the 1980s and that a laser-antimissile system might be possible a decade after that.

Electromagnetism

If a permanent magnet attracts a piece of iron or steel, that is a purely *magnetic* action. If a battery sends electric current through a wire so as to heat it, that is an *electric* effect. But wherever an action takes place involving both magnetism and electricity, such action is said to be *electromagnetic*. There are therefore many manifestations of this phenomenon which was first discovered by the Danish scientist Oersted and greatly enlarged by the subsequent work of the the British physicist Michael Faraday in the first part of the nineteenth century.

One common manifestation of electromagnetism is that a current flowing in a wire produces a magnetic field – this is the operating principle of an electromagnet, and can be harnessed to produce motion in electric motors through the attractive and repulsive forces of magnet fields. When a magnet (either a permanent magnet or electromagnet) is moved near an electrical conductor, turbulent EDDY CURRENTS are induced in the conductor and it experiences a "dragging" force. This dragging force can be used to produce motion, and conversely, the eddy currents can be harnessed to produce a useful electric current (such as in alternators and dynamos). This is an example of a moving magnetic field producing an electric current.

A more complex example of electromagnetism is found in devices such as TRANSFORMERS where a *changing* magnetic field produces a current. Here, two coils of wire are placed close together. When a changing current (changing in amplitude and/or direction) flows through one coil a changing magnetic field is produced, which induces a voltage in the second coil. If this second coil is included in any kind of electric CIRCUIT a current flows.

Understanding by analogy

These phenomena are not fully understood. But in order to exploit them, we devise mental models called *analogs* to help us to obtain at least an

Below: An electromagnet at work. This powerful electromagnetic hoist can lift up to a ton of shredded ferrous material – shown here working in a scrap metal yard.

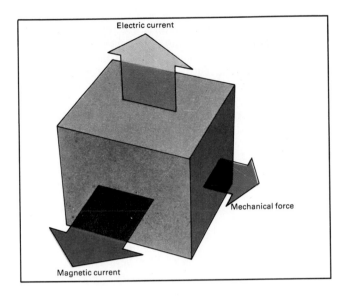

Above: In electromagnetic reactions, the directions in which electric current, magnetic field and mechanical force interact are all at right angles to each other.

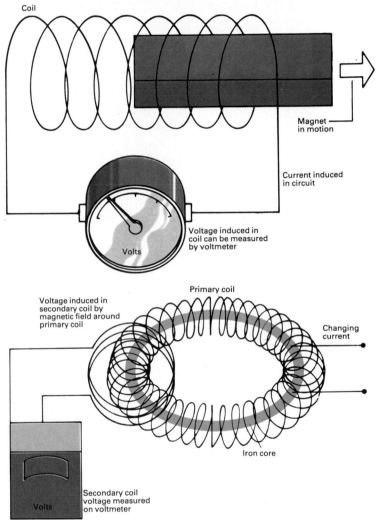

Top: Moving a magnet through a coil of wire induces a current. Above: A transformer consists of two coils wrapped around a ring of iron.

appreciation and a hope that through this means we may learn to design better machines by using a phenomenon which is no more understood than is *gravitation*. It helps to imagine something intangible as something we can see and feel.

For electric circuits we imagine that electrons flow in wires in much the same way that water flows in a pipe. We know that pressure is needed to make water flow so we invent an electrical pressure and call it *electromotive force* (emf) or voltage. The convenience of this analog is that it allows us to use the equivalent of the frictional resistance in the water pipe which increases in proportion to the length of the pipe but decreases in proportion to its cross-sectional area. Then, by another analogy, we can invent a *magnetic circuit*, in which the driving pressure is called *magnetomotive force* (mmf) and the substance which it drives around the circuit is even less real than the flow of electrons in an electric current. We call it *magnetic flux*.

Many authors and teachers declare that, despite its name, flux does not flow. The fact is that it does not exist, except as a human concept, and the only right or wrong about its flow is to be judged on whether the concept is useful to a particular individual. For some, it is more profitable to think of flux as merely being set up because it represents only *stored* energy, and not a continuous loss of power as is the case when electric current flows in a wire. For others, the analog is more profitable if flux is considered to be a more precise analog of electric current so that a magnetic circuit can be given the properties appropriate to those of *inductance* and *capacitance* in an electric circuit.

Linking electric and magnetic circuits

When discussing electric motors, generators and transformers, it is essential to note that each machine includes at least one electric and one magnetic circuit. Since there is no simple equivalent in magnetic circuits to the insulating materials of electric circuits, it is usual to design a machine with only one magnetic circuit but two or more electric circuits. Indeed, the design of magnetic circuits has been likened to attempting to design an electric circuit which must work when immersed in sea water so that although most of the energy flows along the designated paths, an appreciable proportion will follow other routes. For the same reason, electric circuits in machines are usually multiturn coils of relatively thin, insulated wire. Magnetic circuits tend to be single-turn, short and fat.

Above: Faraday's disc dynamo. Contact is made between the copper disc and the horseshoe electromagnet by means of "brushes." When the disc is rotated a voltage is induced in the disc.

The subject of electromagnetism can therefore be expressed as the *linking* of electric and magnetic circuits. In such a linking the driving pressure from one circuit is seen to be derived from the flow in the other, and vice versa. For example, in a transformer an alternating voltage (emf) across the primary windings produces an alternating current in the windings. This produces an alternating mmf in the magnetic circuit, which creates an alternating flux. The alternating flux induces a voltage in the secondary windings, which, if connected in an electrical circuit, produces current.

Vector quantities

The commodity we seek to produce in an electric motor is force which arises as the result of multiplication of flux by current, but it is no ordinary multiplication, for the only quantities of flux and current which are effective are those which cross each other at right angles. Quantities which have both magnitude and direction are called *vectors*, and when determining the interactions of vectors with each other the direction as well as the magnitude must be taken into account. In the above example, the force vector is the result of the *vector multiplication* of the flux and current vectors. Where the flux and current vectors are not at right angles to each other they must be resolved into parallel and right-angular components, but it is always the right-angular components which produce the force vector. Furthermore, the force vector is always at right angles to both the flux and current vectors.

Vector multiplication and, more generally, vector mathematics are only a form of shorthand for handling quantities which have been shown experimentally to interact in this unusual way. This is another example of an analog.

Electromagnetic radiation

The principles of electromagnetism are not limited to electric motor and generator design. Electromagnetic radiation is the name given to a variety of phenomena to which we give different names depending on the context in which we study them. Thus gamma rays, X rays, ultraviolet radiation, visible light, infrared (heat radiation) and wireless (radio) waves are all of the same nature and can all be expressed in terms of a continuous interchange of magnetic and electric energy, each of which pulsates in a plane at right angles to the direction of travel of the radiant waves. All travel at the same speed, about 186,000 miles per second (3×10^8 m/s). The two characteristics which distinguish one kind of radiation from another are the wavelength and frequency. The whole spectrum of radiation extends from very low frequencies with wavelengths of many miles, to incredibly high frequencies of over 10^{22} Hz (1 Hz = 1 cycle/second) and wavelengths less than a millionth of a millionth of an inch.

The study of electromagnetism is therefore basic to the whole of physics, if not to the whole of science. The Earth receives most of its energy from the Sun by electromagnetic radiation. The average private family house in the U.S. contains between 30 and 150 electromagnetic devices (although the higher numbers generally occur where there are several children, each of whom has battery-powered toys). Electromagnetism is basic to the operation of radio and television sets, automobile ignition systems, radar, electric systems, electron microscopes, electric motors and generators, telephones and many other well-known inventions.

See also: Alternator; Dynamo; Electricity; Electric motor; Electromagnetic radiation; Energy storage; Inductance; Magnetism.

Electron

The electron, which is one of the constituent particles of all ATOMS, was discovered at the end of the last century in a series of experiments, the most famous being those of the British physicist J. J. Thomson in 1897. He came to the revolutionary conclusion that "atoms are not indivisible, for negatively electrified particles can be torn from them by the action of electric forces." This dispelled the belief that the atoms were the basic building blocks of all matter.

Thomson had set up an electric field in an evacuated glass tube and detected something coming from the negatively charged electrode (cathode), traveling toward the positively charged electrode (anode) and lighting up the glass of the tube where it struck. It proved possible to bend these cathode rays with magnetic fields and, in this way, to show that they were negatively charged and lighter than any atom (in fact about one two-thousandth of the mass of the hydrogen atom, which is the lightest of the atoms). The principle and the equipment, in more advanced versions, of Thomson's experiment puts the electron at our service today in the CATHODE RAY TUBES of the television set.

The electrons exist in the atom orbiting around the nucleus. They are held there by the electromagnetic attraction existing between the negative charge carried by the electron and the positive charge of the nucleus, just as the Moon is held in orbit by the gravitational attraction between it and the Earth. In the heavier atoms as many as 90 electrons can be swirling in a cloud around the nucleus and, in moving from one orbit to another, they are the source of energy which gives us light and X rays. Also it is the behavior of the electron clouds of the different atoms and the way the clouds link together that gives us the chemical properties of all matter.

Below: Millikan's oil drop experiment, designed to determine the basic charge of an electron. Tiny oil droplets, sprayed from an atomizer between two capacitor plates, acquire a small charge from friction on the nozzle. These drops move under the force of an electric field applied across the plates, and reach a velocity determined by the strength of the field, the electric charge, and other factors. At one point all the forces counteract each other and the drop is suspended. Repeating the process many times gives a series of values for the charge, which are multiples of the charge of a single electron.

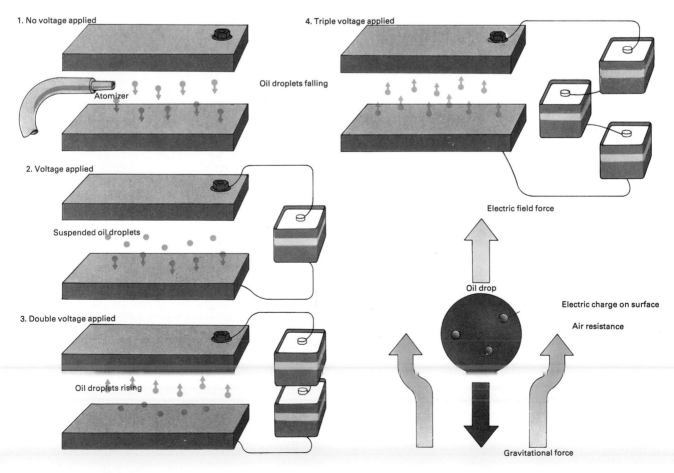

1. No voltage applied

Atomizer

Oil droplets falling

2. Voltage applied

Suspended oil droplets

3. Double voltage applied

Oil droplets rising

4. Triple voltage applied

Electric field force

Oil drop

Electric charge on surface

Air resistance

Gravitational force

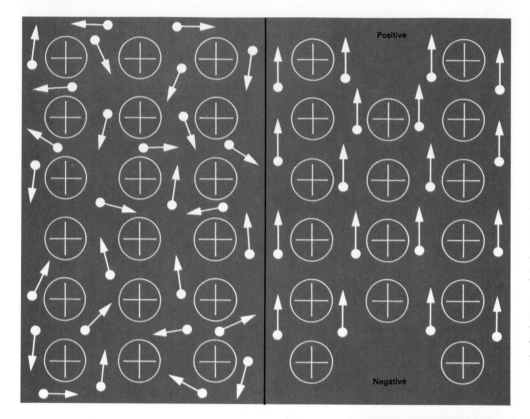

Far left: In a metal, electrons move at random. But when a voltage is applied across the metal the electrons drift to the positive end (near left). This flow constitutes an electric current. If strong enough, it will heat the metal and produce a magnetic field. Below: Bell Labs has crammed 600 microelectronic Josephson devices onto this 0.4 in. chip. These new devices use current flows at superconducting temperatures and work faster than transistors.

The electron is the carrier of electricity, passing with comparative ease through the lattice arrangements of atoms in metals. The unit of electric charge carried by a single electron is, however, very small. A million million million of them are in the electric current flowing through a 100-watt bulb in a second. Other figures which illustrate the small scale of this particle which plays such an important role in our world are: the mass of the electron is one million million million million million millionth of an ounce and its radius is four hundred million millionths of an inch.

When describing semiconductor operation, it is frequently convenient to use the concept of a hole which can be thought of as a place where an electron might be but isn't. This concept is best explained by analogy. If the person at the front of a line of ten people leaves, the subsequent shuffling of those left behind could be described in terms of nine people each moving one place forward. Alternatively, it could be described in terms of a hole starting at the front of the line and moving nine places backward.

One of the most revolutionary developments in physics this century involved the realization that any moving particle can be considered to be a wave and vice versa; a concept called duality. Thus it would theoretically be possible to protect yourself from a bullet traveling toward you by calculating its wavelength and erecting a suitable diffraction grating to deflect it away from you. Unfortunately,

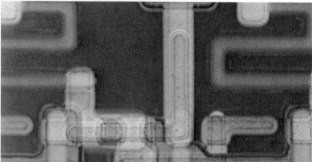

the necessary grating spacing is much less than the physical diameter of any real bullet and so the technique is of no practical use. Electrons have useful properties as both particles and waves. As particles, their interactions with silver halide molecules form the basis of photography and their ability to be deflected by electromagnetic fields forms the basis of picture formation in the cathode ray tube. As waves they can form *diffraction* patterns when passed through foils – like the patterns on the surface of a pond when waves ripple out from dropping in two stones at different places. With the wave-type behavior it is even possible for the electron to pass through barriers which, looking at the electron as a particle, are impenetrable. This is still one of the puzzles of modern physics.

See also: Atom and molecule; Electricity.

Electronic newscaster

The electronic newscaster has seen some changes over recent years. The machine for displaying news bulletins, weather reports and advertisements on large illuminated screens attached to the sides of buildings has changed into a smaller screen, being displayed in many different public places, from restaurants to exhibition centers. Gone are the days where it was the exclusive domain of large public buildings. With the advent of microprocessors and other related new technology, display boards have become within reach of many smaller business establishments.

Old technology
The message to be displayed was first punched onto a long strip of strengthened paper, which was then joined to form a continuous loop. The loop was then loaded onto a machine which read the information from the strip and controlled the lights on the screen. There were usually two of these readers, one for news tapes, and one for advertising. The newscaster read from each one in turn.

The screen
In this kind of machine, the screen is invariably a large incandescent type, consisting of conventional filament lamps. The reader illuminates the lamps in sequence, so that the display moves the message on the screen from right to left.

The small filament lamps forming the screen have traditionally been covered with plastic domes. With four basic colors of lamp, a total of fifteen character colors can be obtained by different combinations of the basic lamps.

New technology
More modern versions of the electronic newscaster are based on microprocessor technology, and are, in effect, small computers. The role of the machine has also changed. Instead of being primarily a news medium with interests in advertising, their role has become almost entirely advertising-oriented. Their role is now mainly to relay local advertising in public places, or to display information.

Hardware
As the machine is now a kind of computer, it easily divides into hardware and software. The display may be one of two types. If it is to be large, then a traditional incandescent type, consisting of light bulbs, is used. For smaller applications, 3 or 6 feet (1 or 2 meter) long displays consisting of LEDs (light emitting diodes) can be used. They have the advantages that they are very much more reliable than filament lamps and consume far less power, but the disadvantage of being just one color – usually red.

The paper tape and readers have been disposed of in favor of computer hardware. Usually a custom-built electronic circuit board is incorporated in the display unit. This has an eight-bit microprocessor and some RAM (random access memory) chips – typical commercial units have approximately 4.5 kilobytes (thousand characters) of storage.

Information to be displayed on the screen is typed in via a typewriter-type keyboard. In doing so the user must interact with the machine's software.

Part of the machine's hardware, as with any computer system, is a clock which is used to synchronize the electronics. The design of the machine usually insures that the hardware clock can be used as a conventional clock, and the current time can be displayed on the screen.

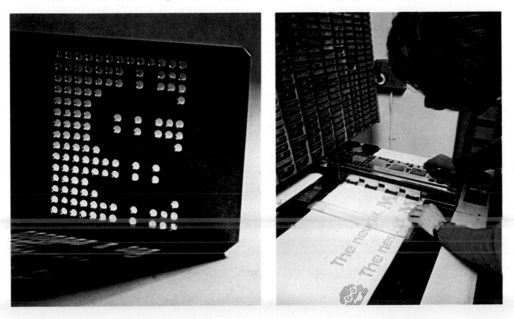

Far left: One of the drilled plates used to control the pattern of the punch dies when preparing the tape. Left: Punching an advertising tape. On this machine the four color tracks are, from top to bottom, blue, green, red and white.

Left: The newscaster in Leicester Square, London. The screens are made of panels 5 ft (1.5 m) high and 4 ft (1.2 m) long. Above: Close-up of the newscaster showing the arrangement of the colored lights.

Sometimes, too, the machines may be equipped with a temperature sensor, so the temperature can be displayed periodically.

Software
The machine's software serves two purposes. It controls the display, and it helps the user enter the messages into the machine.

The software is held in ROM (read-only memory) – actually a chip fixed to the machine's main board. To use the machine, the user simply switches it on and opens a file. A file is simply a batch of information which is stored in the machine's RAM, and may be displayed on the screen. The file may contain an advertisement for a local trader, the weather forecast, news or the specialized needs of the particular location or application. After the files have been created, it is possible to display their contents in any order, by simply programing the machine appropriately. The user can program a series of files to be displayed one after another.

When the machine is switched on, the user may be asked for the time, or the user may opt to set the time if desired. The time feature will use the machine's clock, and the machine may be programed to display the time at appropriate intervals.

Often-used files may be stored on EPROMs (erasable programable read-only memory chips), enabling the message to be used immediately the machine is switched on without typing it in via the keyboard. EPROMs are more popular than the alternative storage medium of cassette tape.

With an electronic system such as this, there are many advantages over the old electrically based system. The machine is very much smaller and cheaper than before. Entering messages is made a lot simpler. There is only one machine involved, and it allows easy editing – mistakes can be rectified immediately without having to retype the entire message. Inexpert typists can now use the system, and achieve results which can be displayed publicly.

The software can give far more control over the display than a conventional paper tape system. Moving messages is just a simple type of computer animation, with the display being treated as a low-resolution computer graphics screen. In addition to the usual right to left movement, the message may be built up in a static position on the display, or it may be slid across the screen from left to right, or from right to left. The possibilities are very wide, needing only suitable software to be written.

Networking
Electronic newscasters are not just confined to one location any more. Many of the newer small displays may be linked together. A centralized control unit can send the same message to the displays via a *network*.

The network may be simply a series of wires connecting a number of displays to the control unit within a single building, or complex of buildings. More ambitious users may opt to site their displays many miles apart. To control the displays simultaneously a modem (MOdulator/DEModulator) must be used to send the information via ordinary telephone wires.

See also: Computer; Computer graphics.

Electronic news gathering

Electronic news gathering – also known as electronic journalism – is a technique that replaces TV news film with electronic pictures. It enables news reporters to get their pictures onto viewers' screens more swiftly than was ever possible before.

The technique evolved in the U.S., where there is fierce competition to be first with the local TV news. Stations have heavily publicized their own investment in electronic news gathering (ENG) with phrases like "Eyewitness News" and "Instantcam."

Live coverage of events – from sporting occasions to embassy sieges – has long been possible via mobile outside broadcast (OB) units. Traditionally, these involve truckloads of equipment and attendant staff. They are slow to reach the scene of action, and not very mobile when they arrive. As a result OB units make only occasional contributions to TV news programs. More mobile, and swifter to react, is the two-person newsfilm crew. But once shot, the film has to be physically transported back to the studio and developed before it can be shown – a very slow process.

ENG looks like the solution. It has the speed of an OB unit and the mobility of the two-person newsfilm crew. In outward appearance, an ENG camera looks similar to a film camera. But it works electronically, in the same way as a normal studio camera. The electronic signals from the camera are recorded on a portable videocassette machine. The recordist in charge of this is also responsible for sound, which occupies its own track on the electronically recorded videotape.

As well as being recorded on a videocassette, the electronic pictures (and sound) can be transmitted instantly back to the TV station. A support vehicle can accompany the ENG crew and beam the news to the studio, either directly, or through a series of relays mounted in line of sight. Thus the pictures can be recorded at base as they are shot or, indeed, transmitted live into a news bulletin.

Camera development

The capacity of two people to record electronic pictures with ENG depends upon a series of technological advances – perhaps the most important of these being in the sheer weight and bulk of the equipment.

In the first portable electronic cameras, an assistant had to walk behind the camera operator carrying some of this equipment in a box – clearly an unsatisfactory arrangement. But a modern ENG camera is a self-contained one-piece unit. All the circuitry is miniaturized, and as automated as possible. Further reductions in weight are being achieved by use of lighter casings, microelectronic components and smaller cathode ray tubes.

Advance in the quality of the pictures that the cameras can produce is also important. Many ENG cameras have facilities for automatic aperture

Left: This very compact self-contained ENG unit is battery-powered, with a two-hour life, and can be carried by one person. A UHF link transmits the pictures directly back to base. An ENG camera works electronically, like a studio camera.
Below: In contrast to the ENG unit, this traditional outside broadcast set-up requires a multitude of staff and much equipment. In some situations, therefore, it is being replaced by ENG.

adjustment in response to changing light, and for minimizing the flare that arises from excessive contrasts of light and shade within a shot. They also contain all the circuitry for coding the color picture ready for recording on cassette. ENG camera tubes tend to be more sensitive than most newsfilm stock, and can operate in lower light levels.

With film, contrast and color balance can be modified after shooting. For example, TV newsfilm can be adjusted when the film is converted into electronic impulses for transmission. But, traditionally, electronic pictures could not easily be altered in that way once they were recorded. The problem was that with an electronic image all the picture information was contained in one combined signal. Now, however, color correctors have been developed which allow the electronically recorded signal to be decoded into its basic components, so the color balance can be adjusted for transmission. This development is valuable, because ENG pictures are often taken in indifferent, fluctuating light conditions beyond the camera operator's control.

Video recorders

Equally essential for the development of ENG was a compact, portable video recorder. The most commonly used type is a refinement of the U-Matic system. This equipment was developed by Sony for use in colleges and businesses and was not intended for broadcasters. But television newscasters in the U.S. spotted the equipment's potential, and pressed the recorders into broadcasting service.

The videotape used is ¾ in. (19 mm) wide – wider than the ½ in. of most home systems but narrower than the 1 in. (25 mm) and 2 in. (50 mm) tapes used elsewhere in broadcasting. Normal U-Matic machines can handle cassettes with up to an hour's playing time. Standard portables, or even more compact models for easier off-the-shoulder operation, can accommodate 20-minute cassettes.

After the broadcasters had found a use for the equipment, Sony upgraded the system. They used the same cassettes, but accommodated more picture information, to offer higher quality. Though these cassettes are the most common format for ENG, some broadcasters contemplate using larger machines with 1 in. (25 mm) tape on open reels.

Even though picture definition is limited by the amount of information that can be crammed onto the small U-Matic tape, this problem can be resolved. There are ways of artificially crispening the picture. More serious is a persistent image jitter that can be visible on such machines. Unadjusted, it could upset all the equipment throughout that chain. Fortunately, the electronics experts have devised an accessory called a *time base correct*. It smooths out all these wrinkles and brings the tapes up to broadcast quality.

Except when pictures are to be inserted live into a news bulletin, program controllers need the ability

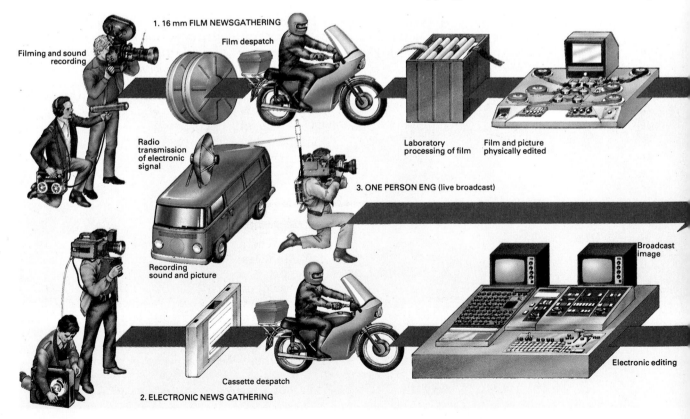

1. 16 mm FILM NEWSGATHERING

Filming and sound recording

Film despatch

Radio transmission of electronic signal

Laboratory processing of film

Film and picture physically edited

3. ONE PERSON ENG (live broadcast)

Recording sound and picture

Broadcast image

Cassette despatch

2. ELECTRONIC NEWS GATHERING

Electronic editing

to *edit* the material – to select which shots and pictures they want to include and which they want to discard. Videotape cannot be edited like film by physically splitting it. Instead, it is edited by re-recording the originally recorded shots, in the desired order, onto a new master tape. A micro-processor-based system controls the operations of the two recorders (the machine playing back the original master tape and the second machine re-recording it). The controller keeps the two machines locked in synchronization, so that a particular edit can be rehearsed and repeated indefinitely. The editor can experiment by adding various new frames onto the last frame of the preceding section on the master tape.

A system called *time code* helps the editing process: it uses a separate track on the tape to record a number which uniquely identifies each frame of picture. This edge numbering system allows an individual frame to be easily found.

In the future

Over the next few years, as ENG equipment spreads even more widely among the world's broadcasters, the technique and its accompanying technology can be expected to undergo further development. For example, there will be continuing advances in the way ENG pictures are rushed back to base. Already, in the U.S., the microwave allocations for TV stations are becoming heavily congested. One solution,

Above: ENG permits live broadcasts, as of the tennis player John McEnroe shown here at Wimbledon.

pioneered by a few U.S. stations, is to transmit their signals across town by infrared light beams.

Many U.S. stations have already begun to make regular use of helicopters to act as a microwave relay station to extend the range over which pictures are transmitted back to base. Sophisticated equipment is required to track the helicopter across the sky, and receive its signals clearly.

During the Vietnam memorial broadcasts a satellite relay station was used. These expensive pieces of technology may be the way forward as geostationary satellites are now ready for use.

Existing camera equipment is already compact enough for solo operation (a separate recording engineer is not necessary) and it is being used in this way by some local U.S. stations. But there will be further development in camera design. In future models the electron tube image sensor will be replaced with a charged-coupled device (CCD). The CCD is a light-sensitive silicon chip, little more than 1/2 in. square. This new system of recording visual images means that the only limiting factor on a camera's size will soon be the lens. For some years, miniature monochrome solid-state cameras have been available for security and defence applications. When such technology can be upgraded to yield broadcast-quality pictures, ENG will be first to benefit.

Technological innovation guarantees ENG a great and developing future. But ultimately the benefits for the viewer – and the viewer's understanding of the world – depend upon the skill and judgment with which TV journalists and technicians use these new tools to gather up-to-date news.

See also: Microchip; Video camera.

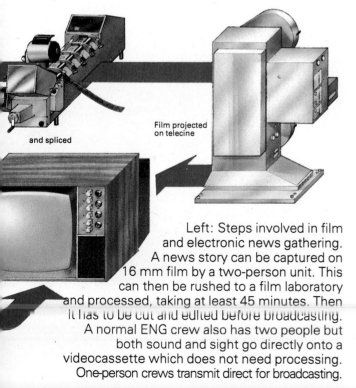

Left: Steps involved in film and electronic news gathering. A news story can be captured on 16 mm film by a two-person unit. This can then be rushed to a film laboratory and processed, taking at least 45 minutes. Then it has to be cut and edited before broadcasting. A normal ENG crew also has two people but both sound and sight go directly onto a videocassette which does not need processing. One-person crews transmit direct for broadcasting.

Film projected on telecine

and spliced

Electronics

Electronics is the branch of science concerned with the understanding, development and applications of electronic devices. On the theoretical level it involves a large number of scientific disciplines including physics, chemistry, THERMODYNAMICS, QUANTUM mechanics and mathematics.

As an applied science it has grown out of electrical engineering, which is concerned with every aspect of electricity: its generation, control, distribution, storage and applications. Today, the electronics industry is an integral part of modern industrial life and is becoming more and more important in the underdeveloped parts of the world as well.

Electronic devices
An electrical circuit is designed to control electric current for a particular purpose and consists of components, or elements, with specific electrical characteristics. The most common components found in electrical circuits are resistors, capacitors, inductors, switches and potentiometers, and they can be used separately or together in a variety of ways. Electronic circuits are a development from electrical circuits and include electronic devices, such as vacuum tubes, gas-filled tubes and SEMICONDUCTORS.

Transducers
Electronic circuits are of little use on their own. To perform useful tasks they must be connected to TRANSDUCERS which convert electric energy into other, more desirable forms. Other types of transducers are used to generate or modify electric signals in response to a wide range of stimuli. For example, in a typical record player, the vibration of the stylus is converted into a small electric signal by the transducer in the pickup cartridge. This signal is then amplified and filtered according to the electric signals received from the volume and tone control transducers before being sent to the loudspeaker, where it is converted back into sound vibrations by another transducer.

The widespread use of electronics today is directly attributable to the wide range of transducers available and the ease with which complicated processing functions can be achieved by electronic components, together with their small size, low cost and high reliability. John Logie Baird used a mechanical scanning system for his original television system while his rivals at EMI used electronic scanning. The Baird system is now history; the EMI system is in worldwide use. Mechanical calculating machines and cash registers have suffered a similar fate. Many other mechanical products from wristwatches to carburetors are being replaced with electronic equivalents.

Thermionic vacuum tubes and related devices
With the advent of the vacuum tube and other related devices, the scope of applications was further increased and new potentials realized. Vacuum tube devices are a particular family of elec-

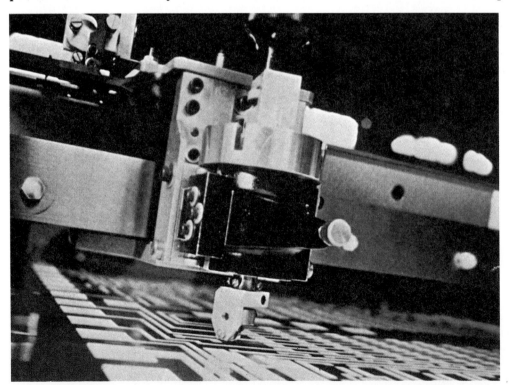

Left: This machine is making a large drawing of the layout of an integrated circuit. The drawing will be reduced photographically to produce a set of masks, which provide the pattern for the engraving of the actual circuit. Light will be shone through the masks to harden the circuit pattern on the silicon wafers.

Right: Semiconductor silicon wafers about to enter a furnace where they are heated to form a silicon dioxide layer. Far right: After heating, the wafers are coated with a chemical called the photo-resist which hardens when exposed to ultra-violet light.

tronic devices and consist basically of two electrodes sealed within an evacuated glass tube. One electrode, called the *cathode*, is constructed from an alkali metal such as cesium which readily boils off electrons when heated. The other electrode, called the *anode*, is usually maintained at a positive potential with respect to the cathode so that the electrons are drawn away and accelerated toward the anode. In this way a current of electrons flows through the device.

This particular vacuum tube is called a DIODE because it contains only two electrodes. These have the property that an electron current can only flow in one direction as no electrons are available at the anode for currents to flow in the other direction, so the diode is used to rectify alternating (AC) currents. This action is similar to that performed by a nonreturn valve in a hydraulic system – hence the name valve used in some countries.

A *triode* is a diode with an extra (third) electrode placed between the anode and cathode. This third electrode is in the form of a mesh or *grid* which allows electrons to pass through relatively unhindered. When, however, a voltage is applied to this grid the electron current flowing through the device is modified or controlled. Small changes in grid voltage can produce large charges in cathode-anode current. Producing a large charge from a small change is the basis of amplification which is essential to the operation of almost every electronic circuit. In some applications two or more grids are required, which can independently control the electron current. Such vacuum tubes are called tetrodes (two grids), pentodes (three grids) and so on.

Tubes were the most common electronic devices during the first half of the twentieth century but have subsequently been replaced by transistors in all but the most specialized applications. Transistors are smaller, lighter, cheaper, longer lived, cooler running, more shock resistant and can easily be powered from simple, low-voltage batteries. They

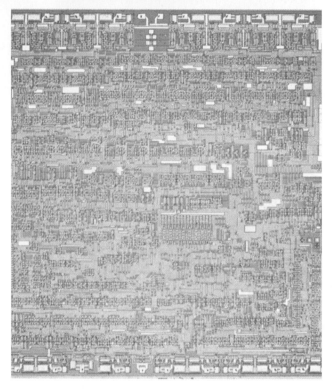

Above: Each 3 in (76 mm) wafer of silicon is made into between 100 and 4000 individual circuits. Planning the layouts is done with a computer.

are normally operational as soon as power is applied and maintain consistent performance throughout their lives.

The CATHODE RAY TUBE (CRT) is a vacuum tube device designed to produce a visual display. Again, electrons are boiled off a hot cathode and drawn toward the (positive) anode, but in the center of this is a hole through which some of the electrons can pass in the form of a beam. At the front of the CRT is a fluorescent screen which glows when bombarded with electrons. Using a grid between anode and

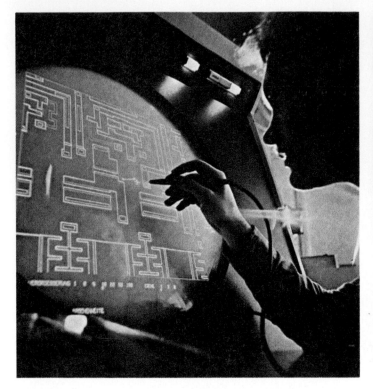

cathode to control the intensity of the beam, a focusing system to focus the beam to a point at the screen, and deflection plates to move the spot on the screen around, a useful display can be created.

The klystron, magnetron and traveling-wave tube are other types of thermionic vacuum tube devices for the amplification of very high frequency signals to high output powers.

Gas-filled tubes

The most common application of gas-filled devices is in light displays such as DISCHARGE TUBES, fluorescent lights, neon lamps and so on. These are, however, not generally described as electronic devices because they do nothing more than provide light. There is no clear distinction here, as, for example, the particular characteristics of a gas discharge make it useful in overload voltage protection equipment. When a certain critical voltage is reached (the *striking* voltage) a discharge is formed and it acts as a low-impedance short circuit, protecting any surrounding components (electronic or otherwise). Also, at a certain magnitude of discharge current these devices exhibit a negative resistance. That is, as the current is increased the voltage between the electrodes drops. This voltage-current characteristic is opposite to a resistor. Such devices can be used in OSCILLATOR circuits.

Gas-filled diodes work on a similar principle to vacuum tube diodes insofar as the cathode is designed to emit electrons. In gas-filled diodes, however, current is conducted through the gas rather

Above left: Planning the layout of a circuit with the aid of a computer. An entire circuit can be simulated without it being built. Above: After the integrated circuits have been built into the silicon wafers, each one is tested on an automatic machine like this one, which scans the circuits through contact probes to test their electronic functions.

than as a beam of electrons. Electrons are knocked out of the cathode by bombardment with positive gas ions (atoms which have lost an electron).

The mercury arc rectifier (diode) is another similar device. Conduction takes place in a pure mercury vapor produced by a spark heating device adjacent to a reservoir of mercury which also forms the cathode. A carbon electrode is commonly used for the anode. Such devices are used where large AC currents need to be rectified.

Other gas-filled devices include the thyratron – a type of triggered diode with three terminals rather like a triode tube – and stepping, or counter, tube devices such as the dekatron. Here, a glow discharge between a cathode and an anode can be made to travel around a numbered tube and so produce a numerical display. This was once a popular method for displaying numbers.

Semiconductor devices

By far the most important area today is semiconductor electronics. But although the semiconductor transistor is a relatively modern invention, semiconducting materials (that is, materials which do not properly conduct but which cannot be classified as insulators) have been known since the beginning of this century.

The *cat's whisker* diode was an early invention consisting of a fine wire contact on the surface of a semiconductor crystal such as carborundum or silicon. It is the junction between the wire and the crystal which produces the diode action.

Copper oxide, selenium and tantalum rectifiers (known as metal rectifiers) can also be classified as semiconducting devices; silicon and germanium semiconductors are the most common electronic devices available today. As pure materials, silicon and germanium make poor conductors (except at elevated temperatures) but when a small amount of impurity is added – called a *donor* – the conducting characteristics change completely.

In its simplest form, the germanium or silicon diode consists of a thin slice of the basic material alloyed on one surface with a donor material (antimony in the case of germanium) to produce an *n-type* semiconductor. This is a material with sufficient electrons in the conduction band (or shell) surrounding each atom nucleus for electron conduction to occur through the material. The other surface is alloyed with another material (for example, indium in the case of germanium) creating a *p-type* semiconductor. This has a deficiency of electrons in the conduction band which can be considered as *holes* or vacancies for electrons and as such are positive. These positive holes can move easily through the material under the influence of an electric field as can electrons in an n-type material.

Between the n-type and p-type sections is the *junction* – a region where neither holes nor electrons can easily move – which acts as a barrier to the flow of current. When the device has a voltage applied across it in one direction, the barrier is reduced and current flows. With the voltage reversed, the barrier is increased and no current flows – this is the diode action.

Perhaps the best-known semiconductor device is the transistor, and there are basically two types. The field effect transistor consists of a bar of either n-type or p-type material called the channel with a connection at each end called the source and drain terminals. The resistance between the source and drain is adjustable by placing an electrode called the gate close to, but insulated from, the channel. If the gate is positively charged, electrons from the channel will be attracted toward the gate and will therefore not be available for carrying current between drain and source. The channel is then said to be depleted. If the gate voltage is reversed, extra electrons are released into the channel, which is then said to be enhanced. The source-drain resistance will then be reduced. It is because of this ability to vary the resistance between two electrodes by varying the voltage at a third electrode that the transistor was so named: trans, meaning across, was combined with resistor.

The junction transistor consists of two diodes back to back forming a p-n-p or n-p-n sandwich. The middle section is called the base, the outer areas are called the collector and emitter. If the base is not connected to an external circuit, very little current will flow between the collector and emitter. When a small current is supplied to the base, a larger current can flow. Both field effect and junction transistors provide amplification and both types can function in linear (analog) circuits and switching (digital) circuits.

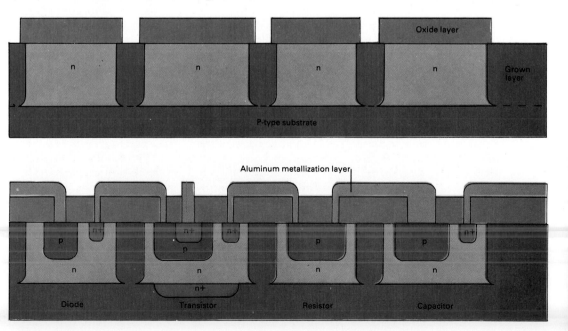

Right: Manufacturing semiconductor integrated circuits. The upper diagram shows the p-type substrate with a layer of n-type material "grown" on top of it. The surface of the n-type layer has been oxidized, and p-type material has diffused into the n-type through gaps cut in the oxide. The lower diagram shows how further diffusions of impurities form various types of semiconductor.

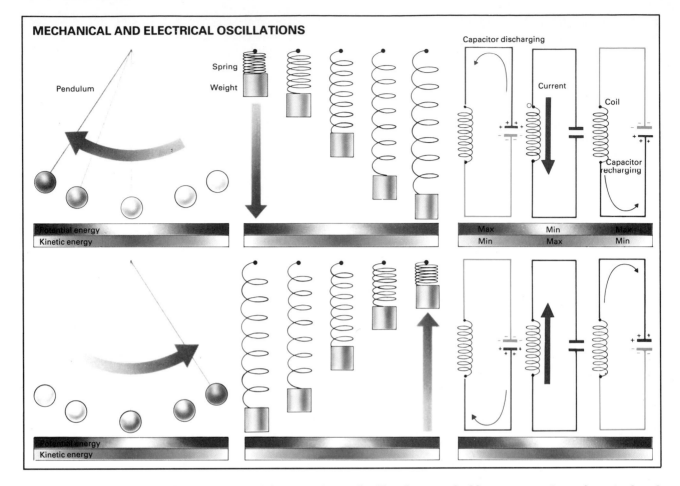

MECHANICAL AND ELECTRICAL OSCILLATIONS

Above: In mechanical oscillations, potential energy is converted into kinetic energy and back to potential energy. In electrical oscillations these two quantities are replaced by voltage and current.

Yet another semiconductor device is the thyristor. The thyristor consists of a p-n-p-n sandwich with three external connections – anode, cathode and gate. This device behaves like a conventional p-n junction diode when it is triggered by a current pulse via the gate terminal connected to the inner p-type section. Thyristors are sometimes called silicon-controlled rectifiers (SCR), behaving in a similar manner to thyratrons, and are particularly useful in power control systems such as electric motor control circuits.

All these devices are available in a variety of shapes and sizes for specific applications and through research and development are constantly being modified and improved.

History of electronics

The study of electricity and electromagnetism flourished with the discovery of a continuous source of electric current – the battery. This occurred in 1800 when Alessandro Volta invented the voltaic pile. For the next half century various electrical and electromagnetic devices were developed – one important invention being the telephone pioneered by A. G. Bell in the 1870s.

The development of electronic equipment was significantly concentrated on communications systems where some means was required for amplifying weak telegraph and telephone signals, and the triode valve proved to be the answer. Lee de Forest announced this discovery in 1906, two years after Sir John Ambrose Fleming had developed the diode. It was not until 1911, however, that the triode was considered as a useful current amplifier. In 1912, feedback circuits were employed in amplifiers and in 1913 feedback was used in an oscillator circuit.

As early as 1887, Heinrich Hertz had demonstrated the physical existence of electromagnetic waves, and in 1895 Guglielmo Marconi demonstrated a primitive form of radio communication (radio telegraphy). For the transmission of speech, however, a source of high-frequency AC voltage is required, which stretched electromechanical systems to their limit. In 1900 a high-frequency alternator was used in transmitters which sent speech signals a distance of 25 miles (40 km). With the introduction of the triode valve driven oscillator, however, the operat-

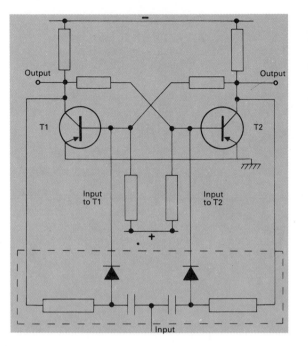

Left: A bistable multivibrator circuit, often called a flip-flop or latch. It has two stable states, T1 on and T2 off, and T1 off and T2 on. The "steering" circuit is within the dotted line. Above: Thick film circuit with discrete capacitors and a transistor.

ing frequencies and transmission distances were increased and in 1916 speech signals were transmitted on medium and short wave frequencies. Regular broadcasting followed in 1920, with successful short-wave radio transmissions over long distances in 1922.

The old crystal set receivers employing the cat's whisker diode flourished during this period, but the cat's whisker device was invented back in 1901. Tube-operated radio receivers became a practical proposition after the development of the high vacuum or *hard* vacuum tube in 1914–15. This improved the characteristics, versatility and lifespan of the tube.

Television was developed during the 1920s following the work of John Logie Baird, although the cathode ray tube that produces the picture was invented back in the late nineteenth century. That radio waves are reflected from the Heaviside-Kennelly layer – a region of the Earth's atmosphere (the ionosphere) – was demonstrated in 1924 and used to determine its height. This experiment led finally to the development of radar, but for this a means of generating very high frequency (microwave) high-power radio waves was required. The magnetron, invented in 1921, proved to be the answer and was first applied in 1935, although it was a subsequent modification – the cavity magnetron – which lead to long-range high-accuracy radar.

In 1948 William Shockley announced the development of the transistor and with this announcement semiconductor electronics came properly into existence. When electronically modifying and processing electrical signals, large currents and voltages are not necessary and can be a disadvan-

tage. The first electronic computer, built during World War II, consisted of more than 18,000 tubes and the heat generated by these presented a major design problem. The size of the transistor, and its low power consumption and dissipation, led to the practical digital computer – one of the most significant of electronic devices.

The need for increasingly compact electronic systems for such applications as space research has lead to microminiaturization and the semiconductor INTEGRATED CIRCUIT. One small slice of semiconductor material – called a *chip* – containing several hundred thousand transistors, enabled great savings to be made in power consumption and removed overheating problems. Because of their size, capacitance and inductance effects could be greatly reduced, thus increasing the maximum operating speeds that could be obtained from these devices.

More recently, metal-oxide semiconductor (MOS) devices operating at very high speeds with minute power consumptions have been developed. These are the modern generation of integrated circuits, not more than a few millimeters in diameter. So small are the operating currents involved that it would be more true to say that they operate on charge transfer.

Semiconductor devices have been slower in development in the field of power electronics because of the problems of removing unwanted heat and the sensitive nature of semiconductor materials to temperature. Heavy-duty rectifiers, thyristors and transistors have, however, been developed.

See also: Electricity; Electron; Physics; Quantum theory; Transducer; Vacuum tube.

Electronics in medicine

In response to the demand for improved knowledge and treatment of illness an increasing amount of electronic apparatus is being used in medicine today. The application of electronics to medicine now involves the use of such devices as amplifiers, stabilized power supplies, LOGIC CIRCUITS (including integrators and differentiators), radio frequency equipment, OSCILLATORS, pulse generators and counters, TRANSDUCERS, OSCILLOSCOPES, ULTRASONIC equipment, recording machines and computers. Electronic apparatus used in medical care and treatment must be electrically safe, reliable, easily serviced, and have good hygienic and ERGONOMIC design. In addition, some components must be able to withstand sterilization. Some types of equipment need to be easily transportable and may need to use internal batteries as a power source. Other types of equipment performing vital functions may need to be connected to an uninterruptible power supply or incorporate failsafe features so that they can never cause danger if they break down. It has been possible to meet most of these requirements since the introduction of the TRANSISTOR and modern solid state devices.

The bulk of electronic equipment is located in hospitals where it is used to assist in the diagnosis and treatment of illness, to monitor the condition of patients, to communicate with and educate hospital staff, to perform administrative services and to control automated processes. Electronic instruments in common use include a wide range of patient monitoring machines which display and record such factors as heart rate, body temperature, blood pressure and brain activity. The information from the monitoring equipment may be displayed at the patient's bedside or at a central nursing station, using a large-screen oscilloscope monitor, pen recorders, and sometimes multichannel tape recorders to record the instrument readings.

Electroencephalographs

The brain generates extremely small electrical currents which, when suitably amplified, produce distinctive traces that can be displayed on an oscilloscope screen or recorded on a pen recorder. The pattern of these brain waves depends on the activity of the brain, which in turn depends on the health of the patient and what he or she is doing. The general rhythms of the waveforms from a healthy brain are fairly consistent from one patient to another, and

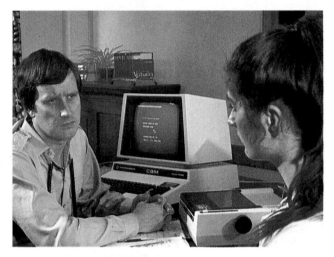

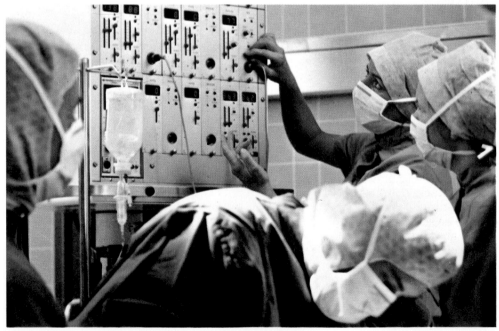

Above: Computers are used in general practice for storing patients' records. They can give information in seconds and preclude the need for filing systems.
Left: This automatic patient surveillance unit can be brought into the operating room. If a critical situation occurs during an operation, it can immediately be recognized.

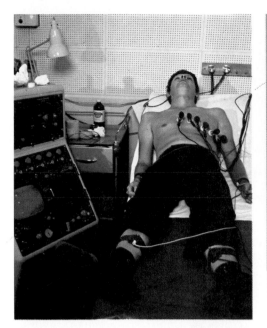

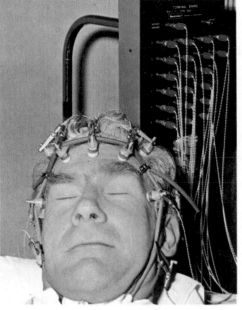

Far left: A patient's heart signals are being monitored by ECG. Disc electrodes are attached to the legs and arms as well as to the chest.
Left: EEG electrodes attached to a patient's head to detect the electric activity of the brain.

any irregularity or abnormality will show up as a distortion of the expected wave pattern.

The electric signals from the brain are very small, typically around 100 microvolts, but they can be detected by electrodes fixed to the scalp or in some cases (such as during brain surgery) placed on the surface of the brain itself. These signals are then fed through high gain amplifiers in an *electroencephalograph* machine whose output signals are used to drive pen recorders or displayed on oscilloscope screens. The electroencephalograph (EEG) machine is widely used both in the diagnosis and detection of brain damage or illness, and in research into the functions of the different parts of the brain.

Electrocardiographs

The electrocardiograph (ECG) machine is related to the EEG machine but its function is the monitoring of the electric signals given off by the muscles of the heart as they pump the blood around the body. When the ECG is used for monitoring the conditions of a patient in the hospital, a set of at least three metal disc electrodes covered on one face with a conductive saline gel are fixed to the patient's chest, gel face down, with adhesive tape. When sample ECG measurements are taken, as in an outpatient department, the electrodes are attached to the chest with rubber suction cups and readings may be taken from the arms and legs as well as the chest. A wire is connected to each electrode and plugged into the ECG amplifier. The signals obtained are in the order of one millivolt, and the *resting rhythm* of the heart is approximately 70 to 80 beats per minute. To display this signal satisfactorily the amplifier has a gain of not less than 1000 and is capable of reproducing exactly frequencies from 0 up to 100 Hz. To

reduce unwanted interference, broad high frequency and very narrow electric power frequency (50 to 60 Hz, depending on the country) filters are included. To protect the patient from any risk of electrocution, the part of the circuit closest to the body (the *buffer* or input amplifiers) is isolated electrically and mechanically from the rest of the circuitry and thus also from the power supply.

Optoisolators are often used to provide electric isolation. These consist of a light emitting diode, which converts an electrical signal into light, placed in a black box with a phototransistor, which converts the light back into an electric signal. Because there is no electric connection between the phototransistor and the light emitting diode, there is no possibility of transmitting an electric shock to the patient. Transformers can also be used to provide isolation. These should incorporate an interwinding screen to prevent capacitive coupling of transient spikes and may need to be designed so that, in normal operation, the core is completely saturated with magnetic flux. This means that a surge of current through the primary will not produce any effect at the secondary, enabling safe use of the ECG.

The signal is displayed on a built-in oscilloscope monitor and can also be used to drive a pen recorder. The ECG apparatus also includes a device which counts the number of heartbeats and a meter to indicate the heart rate. ECG machines may be used to determine the condition of a person's heart during a medical checkup or if heart damage or disease is suspected, and they are also used to monitor the heart activity of a patient in the hospital following a heart attack, accident or serious illness or surgery. In order to detect abnormalities in the ECG rhythm in such cases, the rate meter is fitted with alarm circuits that will trigger an audible or visible alarm to draw the nurse's attention to the change in the patient's condition. To assist the physician in deciding what abnormality has occurred the alarm circuit also triggers a pen recorder which will write out the ECG waveform (this is an advantage in that the pen recorder does not have to be running continuously while the patient is under observation).

Defibrillators

If the heart ceases to function (cardiac arrest) it may be due to *fibrillation*, where the individual muscle fibers of the heart do not contract in a coordinated manner as they should. No characteristic waveform or rhythm can be detected in the ECG, the heart is "shivering" and the patient's circulation is at a standstill. If undetected for more than five minutes this condition will result in the patient's death and so immediate remedial action must be taken by using a *defibrillator*. A portable battery-operated version of this instrument is kept available in hospitals and some ambulances, since a cardiac arrest may occur anywhere at any time.

The purpose of the instrument is to induce the heart to restart its normal beating, and to achieve this two large electrodes are held manually on the chest wall over the heart and a high-energy shock is given to the patient. This has the effect of contracting all the muscles in the chest, including the muscles of the heart, thus restarting the heart action. In some critical conditions it may be necessary to apply the shock several times. The defibrillator contains large-value CAPACITORS that are charged from a stable DC voltage source, which gives them a potential of several kilovolts and an energy content of up to 500 Joules.

The capacitors are charged up, the electrodes applied to the chest wall, and the shock is triggered from a switch on the electrode handles. If the rhythm and waveform of the ECG are not completely absent the defibrillator may be linked to the ECG machine so that the electrical shock is synchronized with the muscular contraction of the patient's heart.

Blood pressure monitors

It is often important to measure the patient's blood pressure, as this gives an indication of the heart's capability to maintain an adequate blood circulation. This is normally done manually, but automated blood pressure monitors are in use in many hospitals. An inflatable rubber cuff is fitted around the patient's upper arm, and inflated to a preset pressure by the monitor to cut off the blood flow in the lower arm. The monitor then actuates the pulse *Korotkoff sound* detectors (these sounds are characteristic of the motion of blood through the main artery in the arm and named after their discoverer)

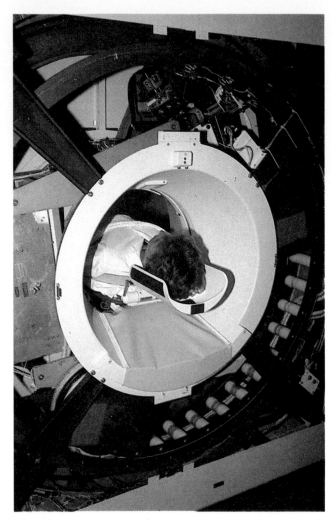

and initiates a controlled leak of air from the cuff. The pressure in the cuff drops until it is equal to the peak pressure in the artery, and at this time the blood is just able to pass underneath the cuff and a pulse can then be detected by the monitor. The pressure reading in the cuff, which corresponds to the peak arterial pressure and is called *systolic* pressure, is stored by the monitor and displayed on a meter. The pressure in the cuff continues to drop until the Korotkoff sounds are detected and again the pressure reading is stored and displayed. This reading corresponds to the minimum or trough pressure and is called the *diastolic* pressure.

Solenoids are used to control inflation and deflation of the cuff. In early designs, a combination of operational amplifiers, discrete transistors and relays was needed but modern systems rely on digital electronics in the form of integrated circuits to perform most of the control functions.

The detection of peak pressure is easily made since it is the first pulse to arrive, but detection of the diastolic point is more difficult and the various types of machine differ basically in how they

CT1010 SCANNER

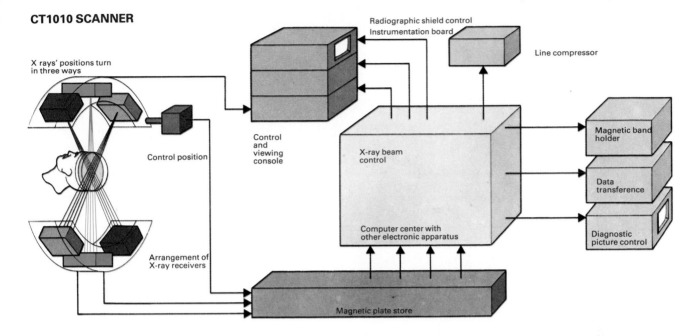

Left: During a Computed Tomography (CT) scan the patient is positioned inside a chamber, with the head held steady. The scanner rotates around the patient, building up a scan of the brain by X rays. After computer analysis, technicians receive a readout of each side of the brain.

Above: A diagram of an EMI surveillance CT1010, which is controlled by a computer. Because of the cost of this machinery, it is out of reach for many hospitals, though its X rays of tissue slices would make early diagnoses possible.

achieve this measurement. The manual, semi and fully automatic cuff equipments all give erroneous readings both at very low (systolic below 80 mm of mercury) and high (systolic above 150 mm of mercury) pressures, so for a patient in a critical condition it may be necessary to use a more direct and accurate method of measuring the blood pressure.

In this method a very fine *catheter* (a nylon tube) is inserted into an artery and connected to a pressure transducer. The dome of the transducer and the catheter itself are kept free from blood by keeping a saline (salt) solution in them, which keeps blood clotting from causing false readings. The transducer has a pressure-sensitive diaphragm, covered by an acrylic plastic dome, which usually has a four-arm strain gauge bridge bonded to it or incorporated in it as in the case of semiconductor strain gauges.

The arterial pressure is displayed on an oscilloscope monitor and can be processed by peak and trough detection circuits to give the systolic and diastolic pressures. In some cases the mean value of the pressure is measured, to determine the mean arterial pressure.

Patient monitoring systems

The ECG and the defibrillator may be used individually, or in the case of patients with more serious or long-term illnesses or injuries, such as those in intensive care units, they may be used in conjunction with other apparatus to provide a continuous monitoring of the patient's condition. Patient monitoring systems may comprise several individual machines separately connected to the patient, or the machines may be physically combined into one main unit such as a multichannel physiological recorder (MCPR). In addition to providing the physician with information on the patient's condition the monitoring systems also contain alarms which alert the nursing staff to any dangerous changes that require urgent attention.

A typical monitoring system might contain instruments to display and record the ECG waveform, the pulse rate and blood pressure, the body temperature and the breathing rate, and some new systems are available which are connected to and supervised by a central computer system.

Other equipment

Modern operating rooms frequently use surgical *diathermy* apparatus. This equipment cuts tissue and coagulates bleeding vessels during surgery by passing a high-frequency current through a small electrode at the site to be treated. The effect is achieved by instantly heating the tissue close to the electrode without affecting the surrounding tissue. Radio frequencies are used since these provide minimal electrical stimulation of muscle tissue. The frequencies chosen fall within a free radiation band. These are bands allocated for uses other than

Left: One of the first electrocardiograph machines made in the 1900s. Wires are attached to metal containers in which the patient put both hands and one foot, recording the electrical changes of the cardiac cycle. Below: Electrodes implanted in the patient's brain stimulate painkilling endorphins.

radio communication and on which interference between users is important. Many operating rooms possess heart-lung and kidney machines, in addition to X-ray, closed circuit television and other equipment.

Outpatient departments use a wide range of electronic equipment, which includes portable diathermy equipment, ECG, EEG, and EMG machines, ultrasonic blood flow measuring apparatus and infrared temperature scanning equipment. The EMG machine (*electromyograph*) is used for investigating the electric activity of stimulated muscle and nerve fibers.

The velocity of a patient's blood flow may be measured by fixing a transducer above a vein or artery, then transmitting an ultrasonic signal through it. The blood will reflect the signal, giving an echo which can be detected. As the blood is moving the frequency of the echo will be different to that of the original signal due to the DOPPLER effect, and by comparing the two frequencies the velocity of the blood flow can be calculated. Ultrasonic echo detection is also used to create pictures of the heart (echocardiography) and to detect brain damage and tumors, using a similar principle to the echo sounding equipment (ASDIC) used by ships. This technique provides a safe method of observing the development of a fetus within its mother's uterus.

Infrared temperature scanning equipment scans an area of a patient's body to provide a picture of the surface temperature variations. An underlying tumor, such as in breast cancer, causes a localized

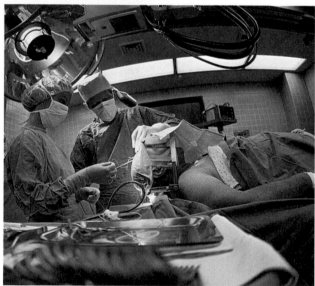

increase in the surface temperature and so can be detected by this technique.

Laboratory support services use many forms of electronic apparatus, including blood typing and clotting equipment, the automated biochemical machines which analyze urine and blood samples and serve to increase the speed with which sample analyses can be made.

In the treatment of certain heart conditions, the patient may be transferred to a unit in the X-ray department where, using image intensifier and closed circuit television techniques, catheters will be guided into the chambers of the heart to observe changes in them and to measure the output of blood from the heart. These observations, together with the results of ECG and other tests, can give the physician a complete picture of the behavior of the patient's heart.

Computers

The digital computer has many uses in the administration of hospitals, controlling such functions as medical records, staff payrolls, stock control and central collation of patient's clinical data. Increasing use is being made of computers in medical research, for example in the detailed analysis of abnormalities in EEG and ECG waveforms.

The body scanner is basically a combination of an X-ray television camera, a computer and an industrial robot. The computer instructs the robot to move the camera around the patient in precise steps while the resultant images are stored in the computer memory. When the computer has built up a complete picture, the stored information may be manipulated and displayed in an appropriate form; for example, a cross-section of the body may be generated showing the exact position of a tumor. This can be compared electronically with readings made on an earlier occasion and any growth areas can be highlighted.

See also: Body scanner; Electronics; Heart-lung machine; Nuclear medicine; Oscillator; Oscilloscope; Solenoid; Transistor; X ray.

• FACT FILE •

- Emil du Bois-Raymond, a physiologist at the University of Berlin, succeeded in 1843 in measuring the electric output of living tissue. He went on to develop, with others, instruments for measuring electricity produced by the human heartbeat. Einthoven's electrocardiograph of 1901 weighed over 600 lb (270 kg).

- Computer-assisted diagnosis is the most controversial application of electronic technology to medicine. In 1961 a medical group in Salt Lake City, Utah, used accumulated data to construct a computer model for aiding the diagnosis of congenital heart disease. Current diagnostic printouts include a statement of how often the computer has made a mistake in a particular diagnosis.

- Radio pills are tiny instruments, fitted with transmitters, small enough to be swallowed, to relay medical information from the stomach and the gut. The first radio pill, devised in Britain, measured internal gut pressure. Later radio pills are sensitive to temperature and pH levels.

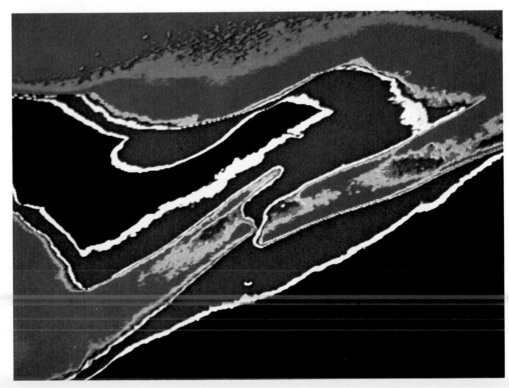

Right: A fracture of the upper arm as revealed by sonography. Sound waves are transmitted through the arm and picked up by the sonograph. Sound intensity reveals tissue density and the signals are changed into a color display — a safer method than X rays.

Electron microscope

The electron microscope was developed to examine specimens in much greater detail than had been previously possible using an ordinary microscope, correctly known as the light microscope. It has proved extremely useful in studying metal as well as biological samples, such as cultures, viruses and cancerous tissue.

In 1873, a German physicist, Ernst Abbe (1840–1905) proved that in order to clearly distinguish between two particles situated closely together, the light source must have a wavelength no more than twice the distance between the particles. This therefore applies to adjacent points in a specimen.

The ability to clearly distinguish two particles is called *resolution*, which should not be confused with magnification. No matter how many times something is magnified, if its image is blurred it will always be so. The wavelength of visible light is approximately 5000 angstrom units, one angstrom being 10^{-8} cm. Thus the minimum details resolvable under a light microscope would be 2500 angstrom units apart.

In the search for a new type of microscope, X rays, which have a much shorter wavelength than ordinary light, were considered, but a lens to control X rays could not be produced. Scientists began to study the electron, finding that accelerated electrons travel with a wave motion similar to that of light but over 100,000 times shorter. Researchers found that either electrostatic or electromagnetic fields could be used to control an electron beam, these lenses behaving in much the same way as the glass lens does in focusing a beam of light. Gradually during the 1930s the electron beam and its control by magnetic lenses were developed to produce shadow pictures of specimens until in 1939 the first commercial electron microscope became available, capable of resolving 24 angstrom units. Since the first commercial instrument, designers have worked toward higher resolutions until today's instruments are capable of resolving 2 angstrom units as a matter of routine.

The electron microscope is an example of accelerated twentieth-century technology reaching a stage of development in less than twenty years, whereas it had taken 300 years to perfect the light microscope.

Transmission electron microscope

The first electron microscope was known as a transmission electron microscope because the electron beam was passed through an ultrathin sample. The variation in density of the specimen resulted in a variation in the brightness of the corresponding area of the shadow image. The transmission electron microscope consists of a vacuum column which is essential for the free passage of the electrons. A tungsten hairpin filament, the cathode, is heated to a point at which it emits electrons. By applying 20,000 to 100,000 volts between the cathode and the anode, the electron beam is accelerated down the column. (This system is called an electron gun.) Condenser lenses control the beam size and brightness before it strikes the specimen, which is mounted on a 3 mm diameter copper mesh grid. The electron beam is focused and magnified by the objective lens before being further magnified and transferred onto a viewing screen by the intermediate and projector lenses. Most instruments cover the magnification range of ×50 to ×800,000. The screen is made of phosphorescent material (zinc phosphide) which glows when struck by the electron beam, and beneath the screen is located a camera for recording the image. Micrographs are not usually recorded at the highest magnification of the particular electron microscope as it is easier to enlarge them later by photographic processes.

The limitation of the transmission electron microscope (as well as the conventional microscope) is that it can focus on only a limited depth of the specimen (depth of field).

Scanning electron microscope

In 1965 a second type of electron microscope became available with a depth of field enabling the study of

Below: A house-dust mite which lives on the sort of debris and general dust visible in the picture. The electron microscope is capable of resolving objects that are much smaller.

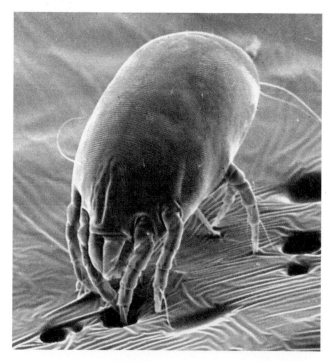

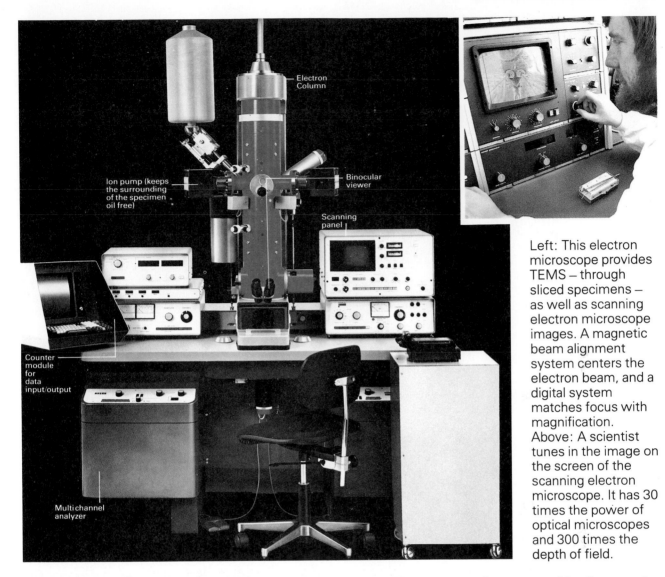

Left: This electron microscope provides TEMS – through sliced specimens – as well as scanning electron microscope images. A magnetic beam alignment system centers the electron beam, and a digital system matches focus with magnification.
Above: A scientist tunes in the image on the screen of the scanning electron microscope. It has 30 times the power of optical microscopes and 300 times the depth of field.

specimens in three dimensions. This new instrument was known as the scanning electron microscope. It employs a column very similar to that of the transmission instrument, consisting of an electron gun and condenser lenses which are used to bounce the electron beam off the surface of the specimen. Situated in the condenser lenses are a pair of coils which deflect a small beam spot across the surface; linked to this scanning system is a CATHODE RAY TUBE (CRT), its electron beam being scanned across the screen in sequence with the beam in the microscope. The electron beam hitting the surface of the specimen drives off secondary electrons, which are drawn toward a detector which, via an amplifier, sends a signal to the grid of the CRT. The greater the number of electrons leaving the specimen the brighter the corresponding spot on the CRT. The magnification of the image depends on the relationship between the size of the area scanned and the size of the CRT, varying between 10 and 200,000

times. The image can be processed by the operator for brightness, contrast and display of either a negative or positive image. A conventional polaroid camera or a roll film camera can be used to record the scanned image.

The smaller the size of the scanned spot the higher the resolution achieved, but as the spot is decreased the energy that it contains decreases. A balance between the energy required to drive off the secondary electrons and the minimum spot size results in a resolution limit of 70 to 100 angstrom units in present-day instruments.

Scanning transmission electron microscope

A third type of microscope first developed in the 1960s and commercially available in 1973 is the scanning transmission electron microscope, or STEM. It combines the most prominent features of its predecessors. The STEM has a new type of electron gun called a *field emission source*. The instru-

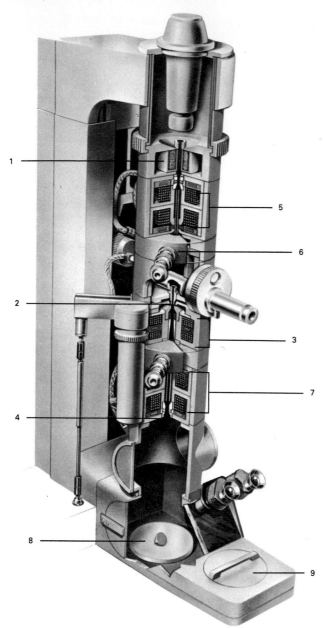

Above: Cutaway view of the Zeiss EM 10 electron microscope shows: 1 Beam alignment; 2 Specimen chamber; 3 Objective lens; 4 Shutter; 5 Double condenser; 6 Coils for adjusting focusing and tilt; 7 Double projector lens; 8 Fluorescent screen; and 9 Specimen camera.

ment scans the electron beam across the specimen, the electrons are collected by a detector and the image is produced through a conventional scanning display system. The field emission source enables a high-energy beam, as fine as a few angstroms in diameter, to be produced. Thus the instrument is able to provide resolution as high as the transmission electron system with the flexibility and image display of the scanning electron microscope.

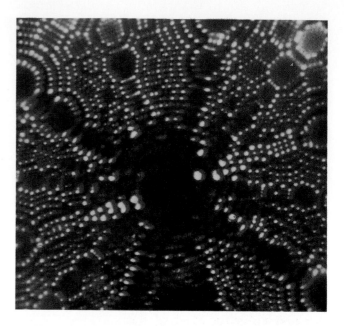

Above: The tip of a tungsten wire seen under the Field-ion microscope, magnification of ×5,000,000.

The largest electron microscope is a three million volt transmission instrument, so large that it is housed in its own building three stories high. This massive instrument produces electrons of enormously high energy, enabling scientists to study specimens many times thicker than those which can be studied in conventional instruments. The ultra-high voltages and the recently developed STEM will enable even more information to be obtained about specimens, using microscopy.

Specimen preparation

In electron microscopy specimen preparation is divided into two categories: transmission, including scanning transmission, and surface scanning. Biological transmission specimens usually undergo a complex preparation before being cut into very thin sections with an expensive instrument called an ultramicrotome; the conventional specimen thickness range extends between 400 and 1000 angstroms. Metallurgical specimens are usually thinned down to less than 1000 angstroms by means of electrochemical polishing. Sometimes metal specimens containing particles can be examined as suitably transparent replica films, formed by deposition onto a thin plastic film that has been previously coated on a specimen grid. On the other hand, specimens for surface scanning are often examined with little or no preparation, but if a sample is nonconducting, a thin layer of gold may be deposited upon its surface to provide good contrast with the scattering of electrons.

See also: Cathode ray tube; Electron; Light.

Electroplating

A wide range of items, from machinery to various household articles, are plated with metal coatings, usually to protect them against CORROSION and also to enhance their appearance, as in chromium plating (which normally involves first depositing a coating of nickel for corrosion protection, followed by a thin overlay of chromium). Sometimes, however, as in the electrogalvanizing of small objects, such as bolts, the function is purely protective. Electroplating may also be used to impart certain other properties to a metal surface, such as hardness, wear resistance and electric or optical qualities.

Coating thicknesses for decorative or protective purposes are generally in the region of 0.001 in. (0.025 mm), whereas greater thicknesses are sometimes required for specific engineering applications.

Principles of electroplating

In electroplating, a metal coating is deposited onto a conducting surface by making the surface metal the cathode in an electrolytic cell, with a suitable electrolyte containing heavy metal IONS – the plating metal. A low-voltage direct current reduces metal ions at the cathode to metal atoms, which adhere to the object being coated, known as the basis metal. The quantity of current is proportional to the weight of metal deposited, according to Faraday's law.

Simultaneously, metal will go into solution at the anode (a sheet or bar of the plating metal), if the anode is soluble. In some instances, however, an insoluble anode is used for practical reasons: lead anodes are used in chromium plating solutions and conduct electric current but remain virtually unaltered – the chromium is removed from the solution and replaced by adding more chromic acid. If the anode is soluble, the weight of metal dissolved from it is proportional to the quantity of current passed. The electrochemical reactions can be shown thus:

Cathode: metal ions + electrons → pure metal
Anode: pure metal − electrons → metal ions

Adhesion

To insure good adhesion of the coating to the basis metal, there must be intimate linkage between the atoms of both metals, and for this reason the surface of the basis metal must be free from scale (caused by the formation of oxides on the metal surface during heat treatment and general production), grease and other deposits. Consequently, thorough preparation of items prior to plating is essential. Preparation may include pickling in dilute hydrochloric or sulfuric acids, mechanical treatment, such as sandblasting, etching, solvent degreasing, or electrochemical cleaning in alkaline solutions.

Throwing power

One of the most important properties of an electroplating solution is its throwing power, which is its ability to deposit a metal coating of uniform thickness on a cathode surface, not all areas of which are equidistant from the anode. Good throwing power enables recessed portions of an article of complicated shape (to which less current penetrates) to be covered with a coating of adequate thickness. This is of particular importance where the basis metal needs to be protected against corrosion. Solutions of ions formed by the combination of a simple metal ion with a neutral molecule normally possess a better throwing power than solutions of simple metal salts.

Commercial electroplating solutions

Commercial electroplating solutions consist of aqueous solutions of heavy metal and other salts to which various specific substances (normally organic compounds) have been added to obtain coatings of

Below: In electroplating, an electric current is passed between the two plates, and metal from the anode passes into solution in the electrolyte to be deposited on the surface of the cathode.

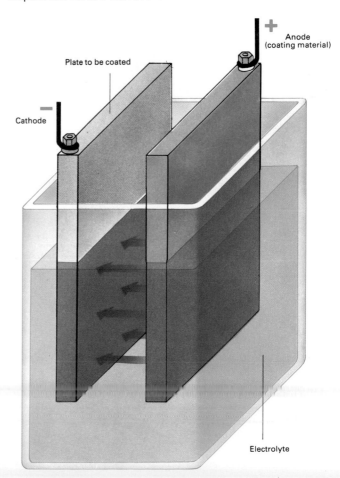

Anode (coating material)

Plate to be coated

Cathode

Electrolyte

the desired properties (for instance, brightness, hardness, ductility, smoothness, and adequate thickness in recess).

Although nonferrous metals and alloys are often electroplated with coatings of various metals, the material most widely used as the basis metal for electroplating is steel. Most metals below aluminum in the *electrochemical series* can be deposited from aqueous plating solutions. The electrochemical or electromotive series is a list of metals in which a metal higher in the series will replace one lower down from a solution of its salts. In order of this series, the main metals are sodium, magesium, aluminum, manganese, zinc, chromium, iron, cobalt, nickel, tin, lead, (hydrogen), copper, mercury, silver, platinum and gold.

Nonaqueous plating solutions using solvents or molten salts for the electrolyte can be used for specialized applications – such as for plating with the more reactive metals like aluminum and magnesium, or for applying coatings to metals like uranium.

In *electroless* plating, the solution decomposes on the surface to be plated (sometimes with the aid of a catalyst) to leave a metal deposit. Although some-times used to plate metals, this process is especially important in the plating of plastics.

A popular plastic for coating is polystyrene which is frequently used for the production of small components, such as automobile door handles and trim. The component is first chemically etched, using a solvent, then cleaned. A surface coating of nickel or copper is deposited by the electroless process. This initial coating is then built up using normal electroplating processes and finished with a chrome plating.

Examples of metals which are used for electroplating commercially are mainly chromium and nickel but also include cadmium, cobalt, copper, gold, iridium, iron, lead, palladium, platinum, rhodium, silver, tin and zinc.

Decorative and protective coatings

Public taste demands that decorative and protective coatings should be bright, such as bright nickel-chromium finishes on automobile bumper bars, hubcaps and door handles, and various domestic fittings. In the past, bright finishes could be obtained only by mechanical polishing at various stages of the plating operation, which necessitated degreasing and cleaning after each stage. This was time-consuming and costly. For this reason, *bright* plating solutions were developed, which contain organic additives to help the plated surface develop a mirror finish.

Plant for electroplating

Electroplating machinery can be automatic, semiautomatic, manual or barrel treated. In the first three categories, the work is treated individually, or at least is individually suspended on jigs. In the barrel process, bulk quantities of small items are plated in a rotating barrel, which is perforated to allow the electrolyte through and which provides the cathode connection.

Automatic plants process articles through the various pretreatment, plating and posttreatment stages. Such plants are used for large-volume production. In semiautomatic baths, the work is transferred manually from bath to bath.

The size of plating vats can range from a few gallons for the electrodeposition of such costly metals as gold, rhodium and platinum, to several thousand gallons for the deposition of nickel and chromium in large automatic plants. The current used depends on the total surface area of the work being processed, so an important factor is the current density – current in amps (A) per unit area. Current density can vary from about 0.013A/sq in. (0.002 A/cm²) for some noble metals, to 3 or more A/sq in. for certain nickel solutions and chromium baths.

See also: **Chromium; Corrosion; Electrolysis; Galvanizing; Nickel.**

Below: Copper plating in a cyanide bath containing cuprous cyanide, sodium cyanide and carbonate, and sodium thiosulfate as a brightener.

Element, chemical

A chemical element is the simplest form of matter; about 90 of them are found in nature, and are numbered according to the number of electrons in their atoms. The heaviest element which is found in worthwhile amounts is uranium (atomic number 92), but in 1972 traces of plutonium (atomic number 94) were discovered. Still heavier ones can be created artificially in PARTICLE ACCELERATORS by firing subatomic particles (protons and neutrons) at a heavy element. Elements as high as atomic number 106 have been observed. Theory predicts that stable, superheavy elements with atomic numbers in the range 110 to 200 could exist, but it has been impossible to create them. The elements that concern us, however, are not the synthesized radioactive ones but those that form the building blocks of all substances. Some elements – usually the fairly inert ones such as gold, platinum, copper, nitrogen – occur free in nature, but most are combined with other elements to form (sometimes highly complex) chemical COMPOUNDS or mixtures of compounds.

An element is in fact made up of the same type of atoms. So a bar of pure metal, say the element copper, would consist of millions of individual copper atoms packed tightly together.

As long ago as the sixth century BC, the Greek philosophers developed theories of matter in terms of primary elements: water, air, fire and earth. It was believed that all known substances could be formed from these elements, either individually or together. It was Robert Boyle who coined the word element for simple, pure substances. Then in 1803 John Dalton published the results of his investigations into the way elements combined in a law of chemical composition. It was therefore possible to find out the relative weights of atoms from the proportions in which they combined in chemical compounds. For instance he found that eight parts by weight of oxygen and one part by weight of hydrogen were needed to produce water.

The chemical elements can be distinguished by their *atomic mass*, the quantity of matter contained in the respective atoms. For example, the lightest of the atomic masses is that of hydrogen with a mass of about 0.04 million, million, million millionth (4 × 10^{-26}) oz. These are rather clumsy units, however, and it is easier to use their ATOMIC WEIGHT which is a number comparing the relative weight with that of the most commonly found ISOTOPE of carbon. Isotopes of any particular atom have a different number of neutrons in the nucleus, and this alters the atomic weight, but chemically this makes no difference. The separation of isotopes relies on the physical differences between them. The atomic weight of carbon is taken as 12 precisely and the other elements are hydrogen, 1.008; helium, 4.003; oxygen, 15.999; uranium, 238.03, and so on. Until 1961 atomic weights were based on taking the oxygen atom as 16 rather than carbon as 12.

Classification

Of the total of 100 or so elements about 65 are classified as metals, 15 as nonmetals and six as INERT GASES. The remainder are not easy to place; these include arsenic, germanium and antimony. Fortunately, the chemical behavior of these elements is not random. Distinct similarities exist, such as in the group of alkali metals, namely, lithium, sodium, potassium, rubidium, and cesium and in other groups, such as the halogens (fluorine, chlorine, bromine, and iodine). When all the elements are listed in a table in order of atomic number, elements with similar characteristics (belonging to the same group) fall at regular intervals. The table is known as the periodic table of the elements.

See also: Atom and molecule; Compound.

Below: The distribution of chemical elements in the planet Earth – the most predominant being spread in mineral rocks and mineral outcrops.

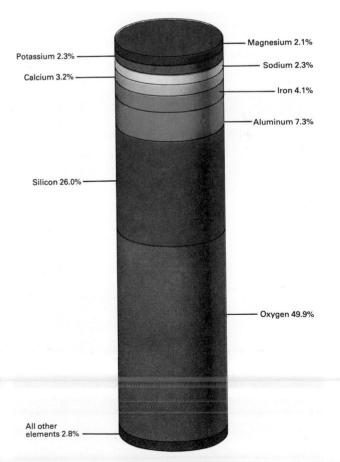

Potassium 2.3%
Calcium 3.2%
Silicon 26.0%
All other elements 2.8%
Magnesium 2.1%
Sodium 2.3%
Iron 4.1%
Aluminum 7.3%
Oxygen 49.9%

Synthetic elements

Most of the material world is made up of only a few chemical elements such as carbon, oxygen, nitrogen, calcium and iron. However, many other elements are known. By the end of the nineteenth century, it was believed that there were 93 different elements, the 93rd being uranium. Beyond uranium, all elements are synthetic. They are, therefore, known as *transuranium elements*.

It was not until 1940 that two Americans, Edwin McMillan and Philip Abelson, managed to produce the first synthetic element — neptunium — by bombarding uranium with neutrons. If a nucleus of the isotope uranium-238 is bombarded with neutrons, neptunium might form in two ways. The nucleus can emit a neutron to give uranium-237, which then undergoes beta decay to form neptunium-237, with 93 protons.

Alternatively, the bombarded nucleus can give off gamma radiation, followed by beta decay. The result is an atom of neptunium-239, which is still unstable. It undergoes beta decay without any further bombardment to produce an atom of plutonium-239, with 94 protons. Thus, transuranium elements just above uranium can be created a step at a time.

When turn-of-the-century scientists postulated that no elements could exist beyond uranium because their nuclei would be unstable, they were not wholly wrong. The synthetic elements are all unstable, but some of their isotopes are long lived.

In terms of the age of the Earth — about 4,500 million years — the half lives of the transuranium elements are short. So, if there were any present when the Earth was formed, they would have changed into other elements long ago.

During the first 20 years of elemental synthesis, a new element was produced on average every two years. Since 1961, when element 103 was first made, only three more elements have been produced with any degree of certainty, although claims have been made for the synthesis of element 107 and element 108.

Making the first six transuranium elements was relatively simple. The next two elements — einsteinium (99) and fermium (100) — have been made in the laboratory, but they were first isolated from the debris of a nuclear explosion that took place in Los Alamos in New Mexico in 1952.

Using a linear accelerator, element 106 is the most recently produced element for which there seems to be good evidence. It has been made by bombarding californium with oxygen nuclei. The resultant element has a half life of less than one second, and its existence can be shown only by studying its decomposition into other elements.

Whereas the type of element to which an atom belongs is determined by the number of protons in its nucleus, its chemical behavior depends on its electrons — how they are arranged and how easily they can escape. Electrons are held around an atom in patterns called shells, each of which can hold a fixed maximum number of electrons. For most elements, these shells are kept distinct by the energy differences between them. Thus, as atoms build up, one shell fills before another starts.

There is one complication: the building-up principle works well until we reach lanthanum, element 57. Lanthanum has three electrons in its outermost shell, but part of one of its inner shells — which can hold 14 electrons — is empty. The next 14 elements, called the lanthanide series, have the same outer shell configuration as lanthanum, but gradually the inner shell fills up. Because chemistry is determined primarily by the number of outer shell electrons, the lanthanides are all very similar chem-

Below: The massive energy needed to synthesize elements is achieved in accelerators. A linear accelerator (shown here) speeds particles through a long straight tube.

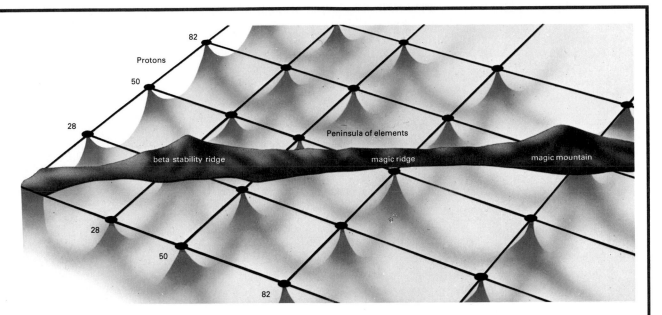

82
Protons
50
28
Peninsula of elements
beta stability ridge
magic ridge
magic mountain
28
50
82

Above: Modern scientific theory predicts the synthesis of superheavy elements, some isotopes of which could be extremely long-lived. On a three-dimensional plot, predicted superheavies form an island of stability beyond the peninsula of known elements.

ically. After the 14th electron goes into the inner shell, the outer shell begins to take on more electrons, so the chemistry changes in step.

This electron mixup occurs again following the 89th element, actinium. In 1944, Glenn Seaborg, who later won a Nobel Prize in chemistry for his work on the synthesis of transuranium elements, predicted that the next 14 elements (up to 103) would form an actinide series (all with similar chemistry). This prediction has been borne out.

Is it possible to synthesize more than the 14 or so known synthetic elements? The difficulty in building up elements one after the other is becoming insuperable. A stage has been reached where the elements are so unstable that it might not be possible to detect their formation. But, if current nuclear theory is correct, although a further small step forward may not be possible a leap might be.

Recent theoretical studies of the structure of the atomic nucleus indicate that it, too, has shells like electron shells. Just as a filled shell of electrons creates stability, so does a filled nuclear shell. The numbers of neutrons or protons that make up a filled shell are called *magic numbers*. And *doubly magic* elements (those that have a magic number of neutrons and a magic number of protons) are particularly stable.

It is possible to picture the known elements as a peninsula surrounded by a sea of instability. The most recently synthesized elements are close to the water's edge – at the tip of the peninsula. Out

beyond, it is possible that there might be an island of stability – a group of elements with atomic numbers around 114, some isotopes of which could have half lives of millions of years.

If an island of stability does exist, and if the elements that make it up were ever present on Earth, there might still be traces of them today. In 1976, scientists from Florida State University thought they had found evidence of such superheavy elements in a mineral sample. The evidence was indirect and not widely accepted. Nevertheless, the search for superheavies continues.

It is difficult to select elements to use as the bombarding nuclei which would lead to a high enough number of neutrons. It might be possible to produce element 114 by bombarding plutonium-244 with calcium nuclei, but no one has yet succeeded in doing so.

Another possibility is to bombard uranium with uranium nuclei in the hope of forming a heavy – but also extremely unstable – nucleus. This nucleus would then break apart by fission into two unequally sized parts, one being a superheavy island-of-stability element. An added attraction of this approach is that it might lead to the discovery of other islands even farther from the end of the peninsula, with stable atoms of even higher atomic number. Glenn Seaborg has calculated that the absolute maximum atomic number, beyond which stability is impossible, may fall somewhere in the range of 170 to 200.

Elevator

An elevator is basically a means of vertical transportation between the floors of a building, comprising an enclosed car balanced by a counterweight and moved by a wire rope.

The earliest elevators were person-powered, using pulleys and a control rope passing through the car, and this system is still found occasionally in some small goods elevators and builders' hoists. The first power elevators were developed in the mid-nineteenth century, using hydraulic power in the form of an extending ram carrying the car and operated by water pressure.

At the turn of the century, it became clear that the future of elevators was to be closely linked with the development of electric power. The first electric elevators were sturdy and simple, and designed for use with direct current (DC) power. The introduction of alternating current (AC) supplies increased design problems, and until the mid-1920s electric elevators used high-speed motors which turned the main drive wheel through a worm reduction gear.

The first practical variable voltage control system comprised a DC hoist motor supplied with power by a motor generator running off the AC supply. Today, geared motors provide the power for most elevators, running at speeds of up to 400 ft (122 m) per minute. At higher speeds, the slow-speed gearless motor has many advantages in both speed of travel and in running costs.

Controls

Although the constant introduction of new materials, such as laminated plastics and stainless steel, have changed the appearance of elevator cars, the basic travel system has changed little during the last 50 years. By contrast, the control system has changed out of all recognition since the early days, when a pull on a rope actuated a pressure valve or moved a sliding bar across the control panel.

Push-button and touch-button controls in the car and at landing stations, automatic acceleration and deceleration and the demand for greater travel speeds has led to today's unified control systems which provide the fastest possible service from a minimum number of cars.

One of the objects of modern design is to provide a minimum average travel time. The simplest system provides service when any landing call button is depressed. The first call registered is the first one answered; on entering the car the passenger presses a button to register the destination floor, and the car

Left: A typical modern elevator system, with glass-sided cars rising up the center of the building. These elevators are designed to add to the architectural appeal of the building itself, not hidden away as they used to be in the building's interior workings. Some modern glass elevators rise on the outside of the building, giving passengers views out over the surrounding area.

is then dispatched to that floor. Other calls are registered in simple "first come, first served" sequence as passengers arrive.

Automatic control

The simplest automatic control system requires only one call button on each floor, no matter how many elevators there may be in the service bank. The approach of a car is signaled by an electrically illuminated arrow (up or down) above the appropriate landing station.

All landing calls are divided into sectors, each comprising a number of adjacent floors. The number of sectors equals the number of cars and a car becomes available to answer a demand for sector service when the doors have closed and there are no assigned calls to be cleared.

As the car becomes available it is allocated to the dispatching system, in which the nearest car to a priority sector is allocated to that sector. The nearest car is chosen by an electronic assembly which constantly compares the location of all the cars in the bank with priority demand.

In many systems, time is saved by initiating the dispatch sequence immediately passengers have stopped entering or leaving the car, a situation detected by photoelectric cells in the door edge.

Most control systems are electromechanical, but an increasing number of elevator installations are controlled by microprocessor systems, which use a complex arrangement of logic circuits to control the movements of the cars.

Counterweights

The cars of passenger elevators are usually counterbalanced by a heavy counterweight equal to the weight of the empty car plus about 45 per cent of its maximum load. The effect of this counterweight is to reduce the amount of power needed to raise the car, and to provide traction between the ropes and drive sheave.

Safety

Passenger safety is an important feature of elevator design. Under normal circumstances, car speed is controlled by *governor* switches acting on the motor and brake circuits. In the event of the speed of a descending car exceeding a predetermined limit, powerful braking arms (activated by a cable connected to a governor unit on the winding machine) are brought into contact with the guide rails to stop the car smoothly and safely. Hydraulic or spring-operated buffers are situated at the bottom of all elevator shafts to safely stop any loaded car traveling at full speed.

In many modern installations, there are devices to sense the weight of the loaded car, and when the car is fully loaded it will bypass all landing calls and

Above: The entrance hall of the Regency Hotel in Atlanta, Georgia. Panoramic cars glide up and down the central column.

service only those calls registered within the car. If the weighing mechanism detects an overload, the starting circuits will not function and the car will remain at rest.

Door interlocks are fitted so that the car will not move until all the doors are fully closed, to prevent the landing doors opening unless a car is present, and to insure that the car doors remain closed until the car has stopped at a landing. Additional devices protect passengers by reopening the doors if they begin to close as a passenger is entering or leaving.

Paternoster elevator

A type of elevator once fairly common but now seldom used is the paternoster elevator, comprising a set of cars carried on a pair of endless drive chains. There are no doors on the cars or on the landing stations, and the cars do not stop but move slowly enough to allow passengers to step in and out. The cars travel up on one side of the chain loop, over the top, and down the other side. Each car is attached to the chains by pivots at its top, so that it always hangs vertically as it passes over the top of the loop or under the bottom.

See also: Electric motor; Governor; Logic circuit.

Elevator, grain

Bulk handled materials such as grain often have to be raised from one level to another using the least possible floor space and in the shortest time possible; this is achieved by the use of elevators.

There are various types of elevator in general use but the principal types are: the bucket, pneumatic, chain and Archimedean screw elevators. They are often used in handling materials other than grain.

Bucket elevator
Bucket elevators are the most common. They are used in a wide range of industries for handling materials as diverse as grain or limestone. Both the lift heights and the capacities of these machines vary greatly, for example, from 10 ft (3 m) lift height at 2 tons per hour, to as much as 200 ft to 300 ft (61 to 91 m) lift height at a rate of 2000 tons per hour in a large bulk grain intake system. It is most essential that the type of elevator used is matched to the material to be handled and this important function is achieved by using either a *centrifugal* or a *positive* discharge unit.

In the centrifugal machines, the material leaves the bucket at a tangent to its path, impelled by natural forces; in the positive discharge machines, the buckets are completely inverted at the discharge point. The buckets are transported by either a belt or a chain. In the centrifugal discharge machines, the use of the belt is more usual; the positive discharge units have chain, or possibly twin chains on the larger units.

Pneumatic elevator
This type of unit is extremely popular in docks for the unloading of ocean-going ships into lighters or shore installations at rates as high as 2000 tons per hour. Such large capacities are needed to insure rapid unloading of bulk cargo ships with loads of up to 60,000 tons. The grain is drawn up through hoses by suction created by a centrifugal fan or rotary blower operating at the discharge end. The operation of the flexible intake hoses on jibs allows this system to effectively solve the problems of trimming the hold and simultaneously accommodating variations in height due to tidal changes. On a smaller scale, in the region of 70 to 80 tons per hour, similar systems are mounted on trailers and powered by an integral diesel engine. This small unit benefits from its mobility and operates just as flexibly as the static unit. Pneumatic elevators may be used with bucket elevators to clean out the bottom of a hold.

Chain-type elevator
This consists of a steel tube enclosing a continuous chain which has extensions or paddles on the links to carry the grain along. Machines of this type operate on the *en masse* principle, which is that once the supply of material to the unit is started it is allowed to fill the elevator from inlet to outlet; thus the particles of the material being conveyed propel each other along in conjunction with the chain of the machine. This can be done effectively either horizontally or vertically, or by using a combination.

Archimedean screw elevator
This type of unit, often referred to as an *auger*, operates on the *en masse* principle and the machine is generally inclined at an angle of 70 to 80° to the horizontal, with a length of approximately 30 ft (9 m). Mounted on a two-wheeled chassis and with capacities of 20 to 30 tons per hour, machines of this style are common in the agricultural industry.

Vertical screw elevators are an extension of the horizontal Archimedean screw conveyer principle and operate on an *en masse* basis, but by a series of variations of the pitch and diameter of the blades they can carry material at a predetermined loading vertically up the machine.

See also: Bulk handling.

Left: The hose of a pneumatic elevator unloading grain from the hold of a ship. Static elevator installations may be part of a silo complex, discharging the grain into bins. Mobile units discharge into mills, silos or vehicles.

Enameling

In enameling, a thin layer of glass is fused to the surface of a metal. Being a glass, enamel has a hard glossy surface which is resistant to corrosion, scratching and staining. These properties led to its widespread use for household goods, such as stoves, bathtubs, and cooking utensils, but today its major applications include finishes on kitchen equipment and buildings. The term enamel (more correctly vitreous or porcelain enamel) should not be confused with the use of this word as applied to high-gloss paints.

The technique of enameling to create jewelry and pictures has its origins in antiquity and was known to the early Egyptians, the Celts, and the Romans, and was developed even more by the Byzantines. Recently there has been a widespread revival of the craft of enameling.

Industrial enameling was first developed commercially in the 1850s in Austria and Germany with the enameling of sheet steel. During the latter half of the nineteenth century mass-produced enamel goods became available, and twentieth-century technology has continued to improve the quality of enameled products.

Preparation of powdered glass

In large-scale industrial applications a continuous smelter is used. The well-mixed raw materials are fed in at one end and the molten glass flows out at the other. It is then cooled by pouring into cold water or onto a cooled metal surface. This glass, called *frit*, is easily ground to small particles in a ball mill. For sheet steel enameling the frit is milled with clay, certain electrolytes and water to provide a stable suspension or slurry. The particle size of a *ground* or base coat enamel can be about 0.003 in. (0.07 mm), but *cover* coats are finer, probably less than 0.002 in. (0.04 mm) in diameter.

The chemical composition of the glass varies according to the job it must do, but it is important that the rates of expansion on heating and contraction on cooling are compatible with those of the metal to be enameled. A typical ground coat, which is the first coat to be applied in sheet steel enameling, might contain 39 to 42 per cent borax, feldspar 19 to 21 per cent, silica 28 to 30 per cent, sodium carbonate 7 to 9 per cent and small amounts of other chemicals including cobalt or sometimes molybdenum oxides to help the enamel to adhere well to the metal. The next layer of enamel is the cover coat

Below: An early example of ninth-century Byzantine cloisonné work – the Hope Beresford Cross. The individual cells for the various colored enamels were made of thin gold strips.

• FACT FILE •

- Very few examples have survived of the fragile *plique a jour* enamels, which were produced mainly between the fourteenth and sixteenth centuries, and were intended to have the translucent appearance of stained glass. Originating in Byzantium (today's Istanbul), the process involved firing translucent enamels into wire patterns fixed to thin metal or metal backings, and then removing the backing to enable a light source, such as a candle, to shine through.

- The Chinese emperor Kiang-hsi (1661–1722) founded enameling factories. The main function of these was to produce incense vessels based on the forms of ancient bronzes. These were then donated to the Buddhist temples he inaugurated around Peking.

- It was thought at one time that enamels found in a tomb in the Kuban region, near the Caucasus, proved that enameling techniques were known in this area as early as the ninth century BC. However, Russian experts have declared at least one of these early enamels to be a more recent forgery, casting doubt on the authenticity of the entire find.

(more than one may be applied) and it may contain 23 to 26 per cent borax, 13 to 15 per cent feldspar, 33 to 36 per cent silica, 9 to 13 per cent sodium carbonate and various other chemicals, including 5 to 7 per cent titanium oxide. Titanium oxide is added for opacity and other opaque materials also used include zirconia, antimony oxide and molybdenum oxide. In fact titanium enamels have excellent covering power. For example, one thin coating about three thousandths of an inch thick will mask a dark colored ground coat. Colored enamels are produced by the addition of metal oxides. For example, iron oxide gives red, cobalt oxide gives blue, and chromium oxide gives a green color.

The enameling process

It is most important that the metal surface is clean before the frit is applied, and the surface should be clean for all subsequent enamel coats. The most common industrial metals are steels (including a very low carbon enameling steel for one-coat white enameling), enameling iron and sometimes cast iron. The sheet metal articles are thoroughly cleaned in a series of baths including detergent, acid, alkali and water for thorough rinsing. To enhance enamel bonding, sometimes a thin film of nickel is plated on.

Most enameling is done by the wet process: a thick slurry of the frit is applied by spraying or dipping followed by draining. The dry process is used for cast iron enameling. Here the first coat is applied wet but the subsequent cover coats are applied by dusting the powdered glass onto the heated article; several dusting and heating cycles are needed for uniformity.

After drying, the enamel is fired either in intermittent box type furnaces or in a continuous tunnel type furnace. In the latter case the articles travel slowly through the furnace on a conveyer, the journey taking about 20 minutes, but they only remain in the hot part of the furnace for about four minutes. Ground coats are normally fired at about

Below: To prepare vitreous enamel frit, the raw materials are thoroughly mixed and fed into a furnace at a temperature of 2360° F (1300° C).

Below: After being coated with enamel slurry, baths travel on an overhead conveyer through a furnace at a temperature of 1560° F (850° C).

1470 to 1560° F (800 to 850° C), while cover coats are fired for shorter times at a slightly lower temperature. A recent development is the use of enamels that can be fired at somewhat lower temperatures – typically about 1020° F (550° C) – and is suitable for enameling aluminum.

On heating, the glass melts and draws up with surface tension to the metal. It is not known exactly why it bonds so well but probably it is a combination of physically gripping the rough metal surface (no matter how smooth a metal may appear it is actually full of minute hills and valleys), and the formation of chemical bonds between the glass and metal.

Jewelry and craft enameling

In many ways the procedures are similar to industrial enameling, but often on an individual scale. The glass formulas differ – a typical craft enamel consisting of 33 per cent silica, 33 per cent red lead, 9 per cent borax, 12 per cent sodium carbonate, 7 per cent potassium nitrate and small percentages of other chemicals. Metallic oxides may be added to the frit as coloring agents, up to 15 per cent. There are three basic types of enamel: transparent, translucent and opaque.

The metals used most frequently are copper or copper alloys, but silver, gold, platinum and stainless steel may also be enameled. The frit is usually applied by dusting it evenly onto the metal surface. If necessary, the surface can be coated with an adhesive, usually gum tragacanth, to make the powdered glass stick. Alternatively the frit may be slurried. There are various design effects in craft enameling: cloisonné, champlevé, and basse taille. In cloisonné the metal surface is divided into individual cells by thin strips of metal which have been soldered or fixed in a colorless flux (glass) to the surface. These cells can then be filled with different colored enamels without any fear of mixing.

The art of cloisonné was developed during the Byzantine period, the cells originally being made of thin gold strips fixed to a gold base. Although it was known to the Romans, champlevé was not widely used until the twelfth century. Cavities are scooped out of the metal to hold the enamel. In basse taille the metal surface is designed in low relief and then covered all over with transparent enamel, so that the design beneath shows through the enamel.

See also: Adhesive; Alloy; Bond, chemical; Ceramics; China manufacture; Heat treatment; Steel manufacture; Surface finishing.

Below: Baths, enameled with a tough white cover, emerge from the furnace. Most white enamels contain titanium oxide.

Above: Unlike glass or pottery, the firing of enamel takes only a few minutes. Here a dish is carefully placed in the muffle furnace.

Endoscope

Endoscopes are a range of instruments for looking inside the human body. A physician can use an endoscope to look for signs of illness. This area of medicine is known as *endoscopy*. The technique of endoscopy has come to most people's attention as an area of high technology. Television pictures from inside the lungs are just one of the more popularly known aspects of this branch of medicine. To look at endoscopy as a modern development, however, would be a mistake, because for more than 100

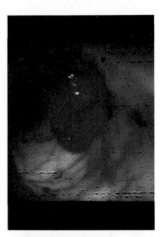

Left and below: A modern endoscope is used to locate and remove a polyp from the patient's colon. This is a common disorder which can cause bleeding from the bowel. The physician's view of the polyp is shown in the upper picture. Using the white handles, the physician removes the polyp with tiny forceps and a snare.

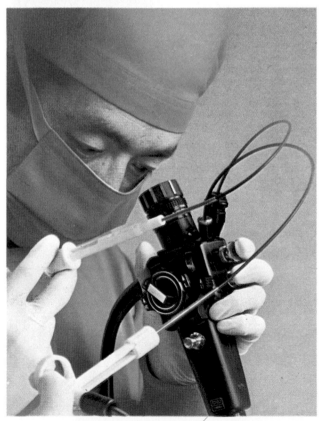

years physicians have been refining the practice of looking inside the human body.

The earliest endoscopes were open tubes illuminated by candles. The physician simply peered down the tube. By the mid-twentieth century, endoscopes had evolved into long, thin telescopes consisting of a rigid tube but containing a series of lenses. At the tip was a miniature light bulb, connected to an external power supply. This kind of endoscope was usually encased in an outer sheath through which saline could flow to insure a clear field of vision.

Fiber optics

The older type of endoscope – which is still used in the treatment of some conditions – was limited by poor light intensity and poor optical definition, owing to the large number of lenses used in the tube. Modern instruments use fiber optics, and are largely free from these disadvantages. Fiber-optic endoscopes consist of thousands of fine glass fibers as thin and flexible as human hair.

Light shining down a single optic fiber either passes straight along it, or, if the fiber is curved, is conducted along its length by *internal reflection*. The fibers' ability for internal reflection is boosted by coating them in a reflective coating. A mixture of two different types of glass is used in each endoscope fiber. The glass comprising the central core of each fiber must have good transmission properties, and the outside glass must have a low refractive index, to limit the amount of light escaping.

In manufacture, a rod of light-transmitting glass is placed within a tube of reflecting glass, and the assembly is heated in a furnace. When the glass is softened, the tube is drawn out into a fiber. The fibers are grouped together into bundles, and encased in a sheath.

When a fiber-optic endoscope is being used to illuminate the insides of a patient, the light source is outside the patient, unlike in the earlier endoscopes. This allows lamps of very high intensity to be used without causing heat damage to the tissues. Typically, halogen lamps are used, although some endoscopes are lit by a xenon arc lamp. With this kind of high-intensity light source, photography (both still and video) becomes possible, rather than simply visual inspection. Often physicians use an instant-picture camera to record the images.

The different arrangement of the fibers within a bundle varies according to the function of the endoscope – illumination or viewing. An endoscope used for viewing or photography has the fibers grouped into *coherent* bundles, in which each fiber occupies the same position (has the same neighbors) at each end of the instrument – this insures that the picture does not become jumbled as it travels along the endoscope. If, on the other hand, the endoscope is to

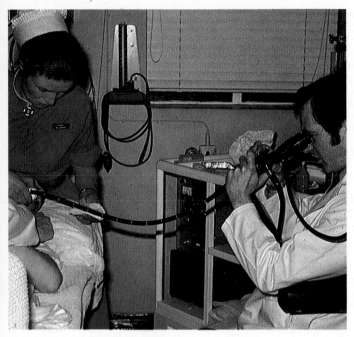

Left: Looking at the interior of a patient's stomach using an endoscope. A teaching attachment enables a second physician to view the images simultaneously. Right: An X-ray picture shows the tortuous path followed by an endoscope shaft during a full examination of the colon. The shaft seen here is about 31 in. (800 mm) long. Below left: A cutaway diagram of a modern endoscope tube, showing bundles of fiber optics. The fibers do not transmit heat rays.

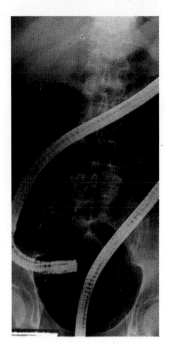

INTERNAL CONSTRUCTION

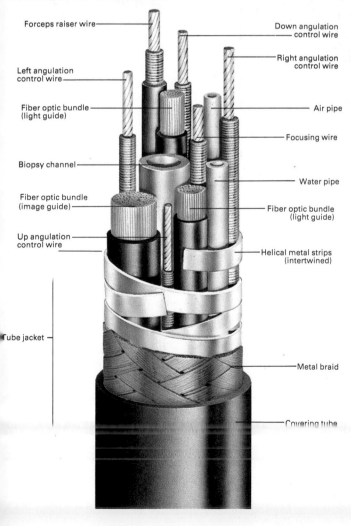

Forceps raiser wire

Left angulation control wire

Fiber optic bundle (light guide)

Biopsy channel

Fiber optic bundle (image guide)

Up angulation control wire

Tube jacket

Down angulation control wire

Right angulation control wire

Air pipe

Focusing wire

Water pipe

Fiber optic bundle (light guide)

Helical metal strips (intertwined)

Metal braid

Covering tube

be used for illumination, the bundle is arranged *incoherently*, or at random, because there is no need to preserve the image of the light source.

Most modern endoscopes are, in fact, used for viewing and illumination, and have at least two fiber-optic bundles. The coherent bundle normally has two lenses: one at the *distal*, or patient's end, which collects the light, and another to bring the image into focus at the eyepiece. There is normally a duct which carries fluid for washing obstacles, such as mucus, from the tip.

At work

A flexible fiber-optic endoscope can be used to look into many hollow parts of the body, including the abdominal cavity, esophagus, rectum, bladder, lungs, stomach, intestines, uterus and knee joints – an endoscope is flexible enough to wriggle around any organ. Endoscopes cannot be used to view the interior of solid organs, such as the liver, but this may change with greater miniaturization.

Although all endoscopes operate on a similar principle, the instrument is given a number of different names according to where in the body it is used. Any particular endoscope may be used for diagnosis, or surgery, as the need arises.

To provide mobility, a wide range of devices can be attached to endoscopes. A kind of pulley is used to thread an endoscope along complicated, curved parts of the body, such as the colon. Some endoscopes have steering mechanisms, and others clearing mechanisms. The larger endoscopes have airways incorporated into them. With an air compressor attached, a physician may distend an organ to improve access for operations or viewing. Other

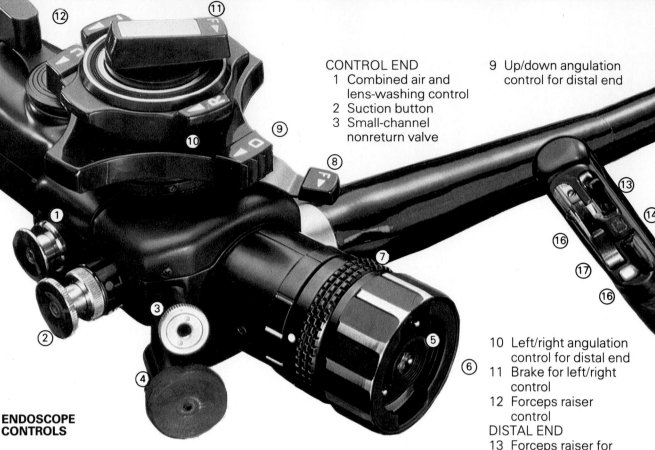

ENDOSCOPE CONTROLS

endoscopes have a vacuum channel for removing fluids. Despite the great complexity of the most advanced endoscopes, they are remarkably compact, usually being about the diameter of a pencil.

Diagnosis

Endoscopes are most often used as a diagnostic tool. More particularly, endoscopy is at its most valuable when it saves a patient from complicated and potentially risky exploratory surgical operations.

In some cases, such as the treatment of peptic ulcers, endoscopy is the only way of certain diagnosis without surgery. Physical examination of the stomach may not help, and X-ray results may be misleading, but a physician can examine the stomach wall by endoscopy and actually see any signs of ulceration.

Endoscopy had been so successful in diagnosis of diseases of the upper gastrointestinal tract that many surgeons use it as first choice, rather than X rays. Endoscopy is also preferred for regular monitoring to detect recurrence of symptoms in many conditions.

Jaundice

The diagnosis of jaundice has been one of endoscopy's successes. The condition is recognizable by a yellow tinge on the skin and the white of the eye. A tiny tube is passed from an endoscope into the patient's biliary and pancreatic systems. A dye is injected so that the biliary and pancreatic ducts show up clearly on X rays. Examining the X ray can show up a gallstone or cancer of the pancreas, either of which may be causing the jaundice.

Surgery

The most advanced kinds of endoscope can be used for surgery, with suitable attachments. Surgery may range from electric cauterization to removal of small pieces of tissue for biopsy. Attachments may range from different kinds of clamps and forceps, to cutting instruments.

Endoscopy can have major advantages over conventional surgical procedures. Removal of polyps in the colon traditionally demanded a complex surgical operation, and a two-week stay in hospital. Now, using an endoscope, the operation is quick, and the patient can return home the next day. The procedure entails using a high-frequency electric current, so the instrument needs to be insulated and the patient is grounded.

Gallstones may be removed with similar equipment. The bile duct is enlarged, and a special basket attached to the endoscope is used to remove

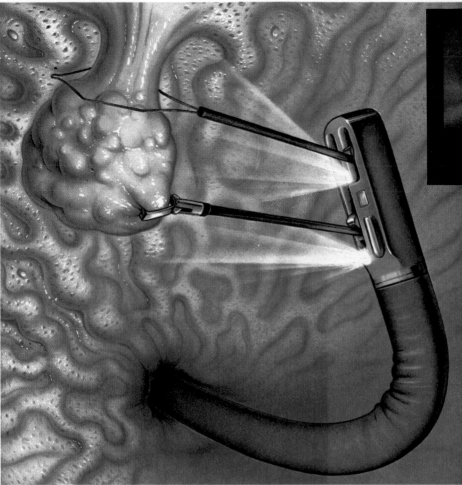

Left: A ployp is removed by means of a side-viewing endoscope. The polyp is snared and a high-frequency current is passed to remove it. It is extracted from the body by forceps.
Above: A surgeon's eye view of the operation, as the roof of the polyp is snared.

the stones. This condition is common in elderly people, who would have difficulty in coping with full-scale surgery.

Endoscopy is well known for its use in laser surgery, particularly to treat major hemorrhaging of the gastrointestinal tract of the elderly. Glass fiber is unsuitable for transmitting laser light, so the endoscope is modified by having a quartz fiber extending along its length.

The site of hemorrhaging is located by the surgeon looking through the endoscope. The tip of the quartz fiber is placed in position over the site, and the laser is triggered, causing the blood to coagulate and stop the hemorrhaging.

Types

Endoscopes are named according to the area of the body in which they are used, rather than the purpose for which they are used. An endoscope used in the stomach is called a *gastroscope*, and it is as thin and flexible as possible so that it can pass easily down the patient's throat. In Japan, gastroscopy is used in mass screening for gastric disease. Gastroscopes can also detect bleeding and tumors.

Another form of endoscopy carried out through the mouth uses a *bronchoscope*. Although a rigid endoscope can be inserted into the trachea, any further penetration into the lungs requires a flexible fiber-optic instrument. Using a controllable tip, the smaller bronchi can be explored to remove objects lodged in the lungs, or even to remove lung cancer or cysts.

If a patient has severe pain or injury in the abdomen, a *laparoscope* may be passed through the wall of the abdomen to locate bleeding or damaged organs. A further use for a laparoscope is to confirm an ectopic pregnancy – a dangerous condition in which the fetus develops outside the womb, in one of the fallopian tubes.

There is a specific endoscope for use in many other parts of the body; they include an *amnioscope*, to check the development of a fetus inside the uterus, a *colonoscope*, to treat tumors and polyps in the colon, and a *ventriculoscope*, to explore the fluid-filled spaces around the brain.

See also: Electronics in medicine; Fiber optics; Instant picture camera; Lens; Microsurgery.

Energy

Energy is the capacity for doing *work* in its broadest scientific sense. A bullet in motion possesses energy because it is in motion and this energy (called *kinetic energy*) is given up or transferred on hitting a target. The energy of the bullet goes into deforming or breaking the target and as heat and sound. A bullet at rest does not possess this energy.

A mass raised against gravitational force possesses *potential energy* because it has a potential to move toward the ground or down hill. If allowed to fall to the ground, this mass can do work – its potential energy can be used, for example, to hammer a pile into the ground.

Kinetic and potential energy are forms of stored energy relating to the motion of bodies or their potential to move and they form the basis of mechanical physics. There are, however, many other forms of energy.

A drum of oil is inert when left to itself, yet it contains latent (undeveloped) chemical energy which can be used in a diesel engine, for example, to drive a truck. The thrust developed by burning liquid fuel in a rocket can impart the energy needed to launch a space vehicle.

Energy is stored in the magnetic field of a permanent magnet because it will move a piece of iron in the vicinity – thus doing work on the iron. Also, energy can be transferred from place to place in the form of radiation. This can be heat radiation or light, in fact, any form of ELECTROMAGNETIC RADIATION. It is in this form that the Earth receives energy from the Sun.

Energy is convertible from one form to another so long as matter is present – in some cases with ease, as in the pendulum which interchanges potential and kinetic energy during its cyclical swinging. In other situations, energy conversion devices are required. In an electric generating station, for example, the chemical energy of coal or oil is released by combustion as heat to raise steam, converted into rotary energy in a turbine, then back into alternating current (AC) electric energy in an electromagnetic generator.

Such conversions are never fully effective. For example, in a power station only about 40 per cent of the latent energy of the fuel is converted into electricity. But if the energy lost – which eventually becomes low-grade heat – is accounted for in the energy balance sheet, the total quantity of energy sums to the same amount on both sides. This is a statement of the principle of energy conservation – one of the CONSERVATION laws in physics.

Below: A solar prominence. The Sun, like all stars, produces enormous quantities of energy from the conversion of a small amount of matter. This leads to constant turmoil in its structure.

Above: Nuclear power plants may help solve the energy crisis. But all fission processes produce dangerous radioactive waste which must be kept isolated for hundreds of years or longer before it decays to a safe level, and storage is a problem.

The nature of energy

It is impossible to say what energy is because, like time, it is a concept so basic that there are no terms available more fundamental to describe it. It can only be quantified in more basic units, for example, by relating it to mass, velocity, temperature and so on. Yet, although the concept is the cornerstone of modern scientific thinking, it is little more than a century old.

Sir Isaac Newton, in formulating his epoch-making LAWS OF MOTION, did not mention energy. The term (from the Greek word meaning work) was coined by Thomas Young (1773–1829) 80 years after Newton and applied to what is now called the kinetic energy of a body. A body of mass m moving with velocity v has a kinetic energy of $\frac{1}{2}mv^2$. Half a century later William Rankine (1820–1872) coined the term potential energy.

Both these terms concern mechanical physics, and their relation to thermal energy heat was not realized until James Joule (1818–1889) demonstrated two crucial experiments. He showed that the heat produced by the passage of an electric current through a wire was related to the square of the current and also that heat was produced by mechanical work. In 1847 he obtained the mechanical equivalent of heat by measuring the temperature rise in water resulting from the action of a paddle driven by a falling mass. Sadi Carnot (1796–1832), father

Above: Using the wind as an energy source demands large numbers of windmills. These are multiblade windpumps with a limited speed. With fewer blades, higher speeds can be reached.

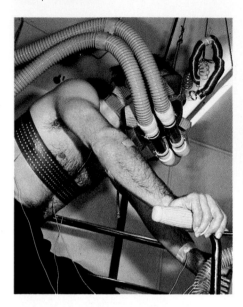

Above: Scientists cautiously take volunteer athletes to their absolute physical limits in controlled experiments and tests.

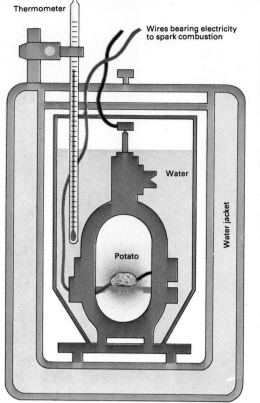

Left: The food we eat provides the energy for our bodies. To measure the calorie content, food is burned and the energy given off recorded by the rise in temperature of the water jacket. The energy came first from the Sun to plants which stored it in chemical form. The animals which eat the plants can release this energy, using it to make the materials their bodies need and to provide themselves with energy for movement.

of the heat engine, never accepted that heat was other than a fluid, called caloric, possessed by hot bodies but this experiment convinced Lord Kelvin (1824–1907). After Joule, the principle of energy convertibility gained rapid ground and with it the concept of energy conservation.

Energy and mass
The law of conservation of energy states that in any system, energy cannot be created or destroyed; this is also kown as the First Law of Thermodynamics. Similarly, the law of conservation of mass (matter) states that in any system, matter cannot be created or destroyed – though in each case the energy or matter can be rearranged. However Albert Einstein (1878–1955) showed that these laws can be combined as the single law of conservation of mass and energy to give the mass-energy equation, which states that in any system the sum of the mass and the energy remains constant. The individual laws are thus approximations which hold true only when the system does not involve nuclear reactions or velocities approaching the speed of light.

Einstein showed that every physical occurrence, of whatever kind, can be specified completely only if it is known when, as well as where, it occurred. A physical event is placed not only in the three dimensions of space but also in the fourth dimension of time. A body in motion therefore exists in such a system and its velocity relative to a (stationary) observer is important in determining the properties of that body. There is a limiting factor here, however, because nobody can travel faster than the speed of light (usually denoted c) and as a body approaches this speed both its observed mass and energy tend to become infinitely large. This led Einstein to the conclusion that a mass m at rest is equivalent to an amount of energy given by m × c², but in motion the effective mass (and therefore the effective energy) increases according to the velocity v of the body in relation to the velocity of light c.

The equivalence of mass and energy has been triumphantly verified. It explains the phenomena of RADIOACTIVITY, the explosion of the A-bomb and the production of energy by nuclear FISSION. Where such a reaction takes place an enormous amount of energy is released accompanied by a reduction (which is actual) in total mass.

Terrestrial energy
Although energy is indestructible there is a tendency in nature for it to become unusable. To be usable (to do work) it must be able to "flow" – to be transferred from place to place. The Second Law of Thermodynamics states that heat cannot be transferred from a cold body to a hotter one by any continuous self-sustaining process. This means that to obtain work heat has to flow from a hot body to a colder one; the energy becomes unusable when there is no heat difference to allow the flow.

As hot bodies become colder and cold bodies become warmer, so gradually a common temperature is reached – the ENTROPY of the system increases to a maximum. Eventually, it would seem, the whole energy content of the universe will become low-grade heat and we will be left with no means of using it. However, this assumes that the universe is a thermodynamically closed system, but this may not be the case.

When the Earth was formed it was endowed with a vast store of energy – the potential energy of its atoms, its internal heat energy and its kinetic energy of rotation. Since then, moreover, it has been receiving energy from the Sun, part of which was stored in the distant past as coal, oil and natural gas, more recently as peat, wood and vegetation. The capital resources, stored in prehistory, can be drawn upon until they are used up (at the risk of altering the atmosphere) and become low-grade heat; additional resources depend on the Sun.

Below: A river provides a useful source of power. The energy available depends on the distance through which water falls. Hydroelectric schemes are most common in hilly areas.

Units of energy
The historic growth, side by side, of branches of physics, between which any interrelation remained unsuspected, led to many different scientific and legal units in which energy can be specified. Examples are calories, therms and British Thermal Units (for heat), watt-hours (for electric energy), foot-pounds and kilogram-meters (for mechanical energy), gauss-oersteds (for the permanent magnet industry) and several others. The modern view of the unit of energy, together with the awkward numerical conversion factors otherwise necessary, led to the adoption of a single basic unit. This is the joule, named after the British scientific amateur who demonstrated energy equivalence in the middle of the nineteenth century.

See also: Energy sources; Energy storage; Matter and energy; Thermodynamics.

Energy sources

Our use of energy has steadily grown during our existence on Earth, from the fires of primitive ancestors to the modern very intensive use in the developed world. Only recently has there been any suggestion that the amount of energy available may not be enough to meet all the requirements. These arise mainly out of the industrial nations' various huge needs, which have not been preplanned but have rather arisen independently of each other. A change in social attitudes, maybe forced by lack of energy, could alter these requirements, just as the discovery of new oil reserves can change the picture as well. It is therefore possible only to give estimates of energy sources and demands, which could turn out to be quite wrong. This article therefore gives current estimates which assume that energy use will carry on in its present form.

One school of opinion believes that it would be preferable to make better use of energy by, for example, insulating homes, using solar energy to run them, and growing food at home.

Energy requirements

The units of energy are joules, or in a more familiar unit, calories. The energy in food is usually quoted in calories or kilocalories. Power stations, however, are usually rated in power units, watts, which do not take into account the time over which the power is used. Energy sources can therefore be quoted in a variety of units, but here the basic power unit of watts (and the multiples kilowatts, kW, and megawatts, MW) will be used.

Note that considerable care is needed when comparing estimates of available energy and energy usage from different sources. The use of different conversion factors and alternative approaches to the estimation of reserves can give significantly different figures. In addition energy sources such as the use of animal dung or wood in underdeveloped countries are almost impossible to quantify. Often energy usage is expressed in units that relate more directly to the production of primary fuels, such as the tons coal equivalent (approx. 8 MWh) and barrels of oil (approx. 10.5 MWh).

Below: The blue glow of Cerenkov radiation, produced when fast electrons travel through water, is a sign of the vast rate of energy production in a nuclear reactor.

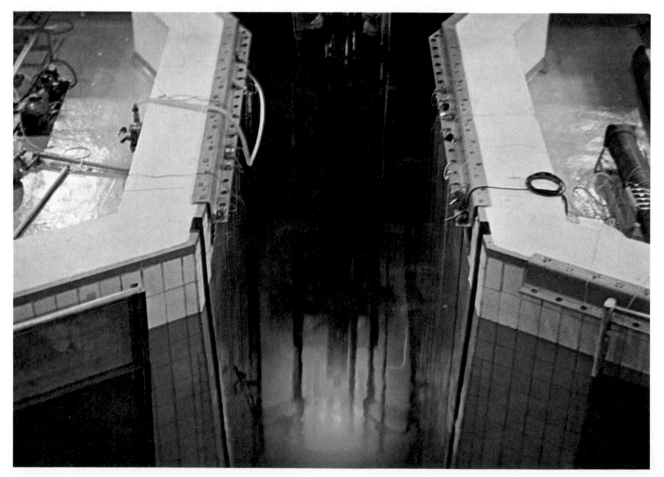

Above and right: A solar chimney, rearing 650 ft (200 m) above Spain's La Mancha plain. Air is drawn under a collector roof at the base and warmed by the Sun, then rises quickly up the chimney. The convection currents can turn a 33 ft (10 m) turbine, giving out 40 kW.

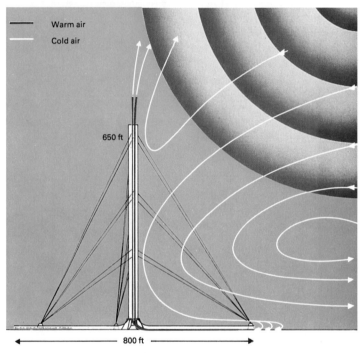

Humans need only 0.15 kW per person as food for light effort (1 kW = 239 calories/second), but Europeans used a total of 5 kW per person in 1970, both for food and for other uses. In the underdeveloped countries the figure was about 0.5 kW, while in the U.S. it was 10 kW.

So far most of the energy has been obtained from the combustion of fossil fuel – coal, oil and natural gas. In the developed countries, the fraction of the energy which is used to produce electricity is estimated to be around ⅓ in the decade 1980–90. The upshot is that the world will require energy at the rate of about 13×10^6 MW, of which 4×10^6 MW will be used for the production of electricity. If the whole world were to be brought up to U.S. levels of consumption these figures would need to be multiplied by six. (1 MW of power = 1340 horsepower; and 1 MW yr of energy is equivalent to one thousand 1 kW fires operating for one year.)

The estimated oil reserves may seem large (430×10^6 MW yr) until the cumulative oil consumption is taken into account: in fact they would be approaching exhaustion around the year 2020 even if the annual increase in consumption had dropped to 5 per cent from 1980 onward, and even if progress is made in developing economic methods of extracting oil from tar sands and oil shales along with the discovery of new fields. In contrast with oil, coal reserves are large; but they will be needed for the production of plastics and oil as well as power. Provided that coal is not used for many more decades as fuel in electricity generating stations (which at present waste at least 60 per cent of the energy), it will be a useful source of hydrocarbons for the chemical industry for another century.

Fossil fuels represent the accumulation of 400 million years of solar energy transformed by *photosynthesis* in plants. On any reasonable timescale they must be regarded as nonrenewable resources, and the end of humanity's brief fossil fuel period of 2000 years is in sight. What are the alternatives?

Water power

Evaporation by the Sun and rainfall on high ground represents one of the largest renewable concentrations of solar energy. The power of water is harnessed by allowing it to fall under gravity through turbines which drive electric generators, and consequently this source of energy is referred to as *hydroelectric power*. The potential world capacity of hydroelectric power is about 2.9×10^6 MW but only 7 per cent is being used. Unfortunately many of the unused sources are far from centers of population and industry, and transmission costs, which are very high, cannot be ignored. Furthermore, such schemes may affect the environment, for example by altering the flow of rivers and causing silting up. The potential of wind energy is also considerable and is being increasingly exploited in favorable areas with "wind farms" containing large numbers of windmills.

Tidal power

The total worldwide potential power of the tide is 3 $\times$ 10⁶ MW, the energy coming from the rotation of the Earth. But the power available from usable shallow seas and estuaries having tidal ranges of more than 10 ft (3 m) is merely 64,000 MW, and the actual electrical output that could be generated is put at 13,000 MW. One tidal barrier in operation is that on the River Rance in France, producing electricity at an average power of 100 MW – compared with a large power station's 1000 MW. Tidal power can make only a minor contribution to world requirements, mainly by improving the overall efficiency of a regional electricity supply in conjunction with pumped storage systems.

Geothermal power

Steam is available from hot springs in volcanic regions and the total installed generating capacity is about 1200 MW. Assuming that about 1 per cent of the potential energy available can be tapped, and converted to electricity with an efficiency of 25 per cent, the potential yield is estimated to be 3 $\times$ 10⁶ MW yr. If withdrawn over a 50-year period, this source would provide about 60,000 MW of power. This source of energy is highly significant to a country like Iceland which has no fossil fuel.

Another possible source of thermal energy is the temperature gradient of the sea. In some areas the surface temperature heated by the Sun is at around 77° F (25° C) while the temperature 3300 ft (1000 m) down is 41° F (5° C) with the difference being sufficient to drive a heat engine.

Solar radiation

About 30 per cent of the Sun's radiation is reflected by the Earth's atmosphere, 20 per cent is absorbed by it, and 50 per cent reaches the Earth's surface. In favorable regions the average intensity at the surface is 600 MW per square kilometer over a nine-hour day. This is certainly a valuable and under-used resource – an area of around 25,000 acres (100 km²) receives an energy input that is equivalent to the total present world usage – but is not easy to exploit. With present technology only around 15 per cent of the solar power in an area can be converted to electricity so large collector areas are required.

An alternative involves collecting the solar power by satellite in space, where the intensity is a con-

stant 1390 MW/km². The energy would be beamed as microwave power to collectors on Earth. These would cover a large area – some 3 square miles (8 km²) for a 1000 MW plant – but would consist of wire arrays which would not prevent the use of land beneath them. Although initial costs would be high, because a Space Shuttle program would be required, the scheme would be nonpolluting with an inexhaustible fuel supply.

Nuclear energy

Nuclear reactors use radioactive fuel to produce heat by FISSION which is then used to make electricity in a more or less conventional way. The initial fuel, uranium-235, is in short supply, but breeder reactors which are being developed actually produce more fissile material than they use.

It may thus appear that nuclear fission is the answer, but a major drawback is the amount of radioactive waste produced. This takes several hundred years to decay to a safe level, and is being produced in increasing amounts. At present it can only be stored. The safe transport of fuel elements is another serious problem.

FUSION, the other means of producing nuclear energy, would be much safer, using deuterium (from sea water) and tritium (from lithium, which is plentiful). So far, however, no practical methods of controlling fusion reactions have been developed.

Below: Wood-burning stoves and fires have come a long way since nomads and tribesmen first used them to cook food and warm themselves.

Vegetation as a power source

The world's annual forest woody increment is estimated to be $12,900 \times 10^6$ tons of which only 13 per cent is harvested. The remaining 87 per cent is capable of producing 5×10^6 MW yr, a quantity approaching the present annual world power consumption. Unfortunately, this source, like the wind, suffers from lack of concentration. The ecological and climatic effects of indiscriminate cutting would be severe but, given scientific management on a continuous basis, forests and other vegetation could be a useful source of power for some underdeveloped countries.

Considerable effort is being applied to the conversion of biomass into more readily usable fuels such as alcohol for vehicular use. Research is also proceeding on the use of photosynthesis processes to directly split water into hydrogen and oxygen gases which can be used as fuels.

The world population is expected to increase about 6×10^9 by the year 2000, and the energy required for food production will then be about 1×10^6 MW. If the energy problem is not solved, it could

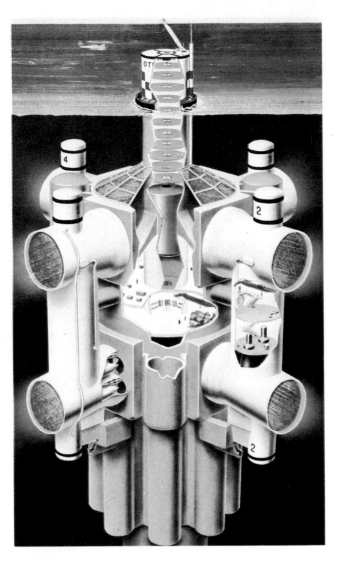

Above: Cutaway of Lockheed's OTEC float, turning hot water into electricity. Hot water is drawn in through the grill vaporizing ammonia in the core to drive a turbine. The ammonia is condensed by cold water from the sea bed.

mean a drop in material standards in the developed countries, or famine in the underdeveloped.

The problems involved in meeting the world's energy requirements are not all technical – just as complex are the social and political aspects. There are, however, many known ways of increasing the efficiency with which we use our energy resources. Some experts believe a reasonable annual increase in demand could be met without a large expansion of fission power, with its attendant radioactive waste problems, and without depleting our resources of fossil hydrocarbons.

See also: Energy; Hydroelectric power; Solar energy; Windmill and windpump.

• FACT FILE •

- The ancient Chinese used natural gas as a fuel before 1000 BC, drilling deep through rock to release the gas, which they then transported through pipelines constructed from bamboo. The gas was used for lighting as well as heating and cooking. The coal mined at the same period was used as a supplement to the gas.

- Asphalt was one of the earliest energy sources, being burned as a fuel in the Middle East as early as 6000 BC. In Mesopotamia, asphalt was used to fuel furnaces for bricks and pottery up to the sixth century BC, after which, despite the existence of huge surface deposits, its use almost disappeared.

- Liquid wastes from industrial processes have become an important addition to the catalog of energy sources. Wastes, such as petroleum and solvent derivatives, and even heavy sludges, can be treated so that they can be atomized into small droplets which can then be injected into a combustion chamber in the presence of oxygen. A Philadelphia plant imports three million gallons (10 million liters) of waste per annum to burn for steam production.

Power plants in space

Solar energy ought, in theory, to be the ideal energy source. It is clean – it produces no pollution, unlike coal and oil, and there are none of the radioactive wastes that make nuclear power so unpopular. It is endless (at least until the Sun burns out many millions of years hence), so it will still be abundant when fossil fuel reserves run low, next century. And solar energy is free – apart from the cost of harnessing it. Making sunlight work efficiently and cheaply is the only problem that has to be solved before solar energy becomes a reality.

Power satellites score over ground-based solar collectors on several counts, most importantly because (above the filtering blanket of our atmosphere) a given collector area can intercept at least four times as much solar energy as the sunniest spot on Earth. In space, there are no clouds, bad weather, or nights to hinder production.

The Solar Power Satellite (SPS) is placed in so-called *geostationary* orbit 22,400 miles (36,000 km) above the Earth's equator – the same type of orbit as used for communications satellites. At this altitude, the satellites orbit at the same rate as the Earth spins, and so seem to hang fixed above a point on the equator. Here, the SPS intercepts sunlight, converts it into *microwaves* (short-wavelength radio waves) and beams them back to collector arrays on Earth – where they can be converted into electricity, with high efficiency.

Depending on its size, each SPS could deliver thousands of millions of watts virtually continuously. In their orbits high above the equator, the satellites are eclipsed by the Earth's shadow only around the equinoxes (late March and September) for up to 72 minutes at a time. But the eclipses occur around local midnight, when demand is usually low and these predictable interruptions can be accommodated within normal electricity load management.

Solar power satellites, however, are unlikely to supplant ground-based power stations completely. More likely, they would supply baseload power, which could be supplemented, if necessary, by terrestrial generating systems. In 1980, a joint study by NASA and the U.S. Department of Energy concluded that it was feasible to construct a fleet of 60 solar power satellites, the first of which would be in operation in 2010 and the last by 2040. The European Space Agency has estimated that 40 such satellites would be needed to supply a quarter of

Above: Proposed geostationary power satellite. Below: One of the arguments against beaming power down to Earth was that the microwave beam might damage human beings. Using a beam stronger at the center reduces the problem but it must be accurate.

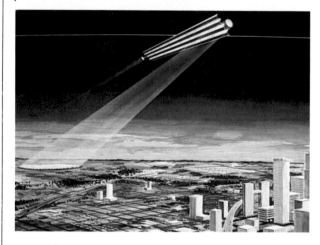

the European Community's electricity requirements in about the year 2030.

There are two main ways in which an SPS can generate power. One way is by focusing sunlight with large concave mirrors made of reflective plastic film to heat helium gas to a temperature of 2550° F (1400° C). The hot helium then turns a turbine, which produces electricity in similar fashion to power stations on Earth. According to one design, 16 such turbogenerators, each delivering 300 MW, are fed with sunlight by one large reflector nearly 3 miles (5 km) across. An interlinked system of four reflector units and associated turbines, stretching a

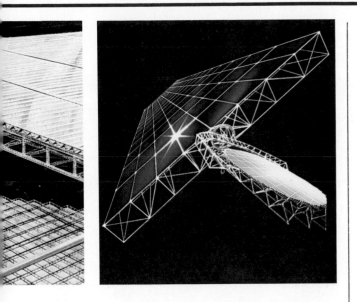

Above: To build a satellite like this – 8 miles (12 km) long and 20,000 tons – needs a ferry with a 200-ton payload and 500 workers for six months. 5,000 MW of electricity is the lure.

total of 11 miles (18 km) in length, could deliver 10 GW to Earth – the equivalent of ten terrestrial power generating stations. Such an orbiting power station is projected to have a mass of 100,000 tons.

An SPS using solar cells is a more popular design because there are no moving parts to go wrong, and the use of solar cells in space to provide electricity for satellites is well established. Solar cells convert sunlight directly into electricity. Usually, they are made of silicon, which is an abundant element in rocks, although other materials such as gallium arsenide may provide greater efficiencies.

Panels, several on a side, are stubbed with solar cells. To deliver 5 GW, an SPS may require 10,000 million cells, covering a massive area of 20 square miles (50 square km). Transferring the solar collectors to space avoids using up vast areas of land that would be required by ground-based collectors of similar output. Remote-controlled robots could be used to construct the lightweight beams which support the blanket of solar cells. Everything in space is weightless, but the structure of the SPS employing solar cells would have a mass of about 50,000 tons.

Whether the SPS uses turbines or solar cells, the electricity generated is converted into microwaves by devices known as *amplitrons* or *klystrons* and is then beamed to Earth through an aerial 1 mile or so in diameter. A wavelength of approximately 10 cm (2450 MHz frequency) is chosen because this type of microwave radiation passes through the atmo-

sphere virtually unabsorbed. At the ground, arrays termed *rectennas* collect the microwaves and convert them very efficiently into electricity. In 1975, microwaves were beamed from the 85 ft (26 m) diameter radio telescope of the Jet Propulsion Laboratory, California, to light a bank of lamps 1 mile (1.5 km) away – the conversion efficiency achieved at the rectenna was 83 per cent. The high efficiency of power transmission and reception is crucial to the economics of placing large solar collectors in orbit.

To prevent the power satellite's beam straying off station either accidentally or by sabotage, the ground array transmits a reference signal to which the satellite locks on. Should this guide signal be lost, a failsafe device insures that the microwave beam spreads out harmlessly. Any realistic assessment of the dangers of power satellites must be balanced against the pollution from fossil fuels, and waste from nuclear reactors.

Receiving arrays for the SPS beam consist of panels studded with T-shaped aerials linked to rectifying devices known as Schottky barrier diodes, which convert the microwaves into electricity. The electricity is then fed into local power grids. They would be best placed in desert areas where land is cheap, but these areas are inevitably farthest from the main centers of consumption. In areas where demand is highest, such as the northeastern seaboard of the U.S. and western Europe, rectennas would need to be placed offshore – but then cost and complexity also increase. The main problem with the SPS, however, is not the health danger of the microwave beam, but radio noise and atmospheric effects.

Although only a small fraction of the incoming power beam is absorbed by the atmosphere, the volume of power being delivered is so great that this fraction is enough to cause considerable radiofrequency interference and possibly heating of the ionosphere. Scattering of the incoming beam by the ionosphere and raindrops could produce noise over a wide range of frequencies. This, in turn, would interfere with many other waves, such as radar, and would completely swamp the sensitive radio telescopes used by astronomers.

Power beams piercing the atmosphere around the world could heat the ionosphere sufficiently to disturb its normal radio-transmitting properties. Optical astronomers worry about the effect of a fleet of power satellites ranged around the equator. Each satellite would appear more brilliant than the brightest star, and a complete ring of them would mean that the sky would never be completely dark, rendering telescopes useless for many purposes.

Energy storage

Energy cannot be created or destroyed, but it can be converted from one form to another. Apart from solar and nuclear energy, the original source of energy is invariably some natural form in which energy was stored and storage of the original resource, such as carrying fuel in the tank of a vehicle, is one form of storage very widely used. There are, however, a number of cases where, having released the energy from its natural storage, it is advantageous to store it again for later recovery, even if this involves a further conversion process and loss of some of the potential of the energy for doing useful work.

The reasons for such storage are: to even out fluctuations in demand; to store energy which might otherwise be wasted so that it can be used later; to accumulate energy over a long period for release very rapidly; and to convey energy to where it is required when continuous transmission systems are not practical.

Pumped-storage systems

The only large-scale method of storing energy which has been developed to operate with the main electric supply system is known as *pumped storage*. A large electric machine which can be used either as a motor or a generator is connected to a water turbine and a water pump. This combination of machines is positioned close to a large water reservoir and connected by a series of pipes to a second reservoir placed at a much higher level. The *head* (difference in level) may vary from 100 to 2300 ft (30 to 700 m) or occasionally even more. Such arrangements are normally found in mountainous country. During off-peak times (usually the night) when there is plenty of electrical generating capacity available elsewhere which it would be uneconomical to shut down, power is supplied to the pumped storage station and used to pump water from the lower reservoir to the high reservoir. At times of high demand (the daytime peak load periods) the pump is disconnected and water is allowed to flow back downhill through the turbine, driving the electrical machine as a generator. The flow of water can be very quickly regulated so that the machine will meet almost instantaneously any demand placed on it up to its maximum capacity.

It is normal for this type of system to generate power for a total of four or five hours each day and to pump up the corresponding amount of water during a period of about six hours during the night. There are inevitable losses in all the machines concerned and friction losses in the pipes, so the total amount of energy recovered is approximately 70 per cent of that absorbed during the pumping operation.

Compressed-air and gas systems

Compressors are frequently used to compress air into large pressure vessels from which the energy can be drawn as demand arises. Such systems are very commonly used to operate large numbers of small pneumatic tools and other equipment in factories. To provide individual electricity-driven tools would be very expensive and inefficient in terms of energy consumption. The compressed air system allows high-efficiency motor-driven compressor systems to be used and also gives a steady electric power demand even if the individual tools present a fluctuating demand. The overall efficiencies for this sort of system are around 50 per cent but the other advantages outweigh the losses.

Compression of gas is also a method of storing up energy in both pneumatic and hydraulic control and power systems. In these systems the pneumatic or hydraulic power may be used to operate parts of machines, to open and close doors on trains, buses and so on. Pneumatic systems use essentially the

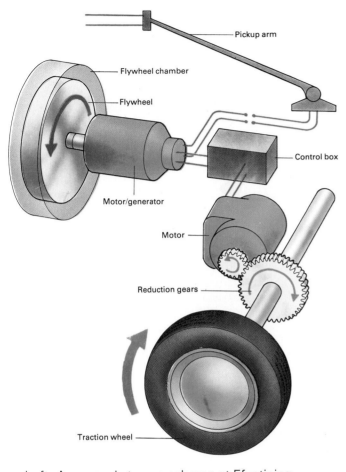

- Pickup arm
- Flywheel chamber
- Flywheel
- Control box
- Motor/generator
- Motor
- Reduction gears
- Traction wheel

Left: A pumped storage scheme at Ffestiniog, Wales. The upper reservoir is 1000 ft (300 m) above the lower plant. Above: The operating principles of a gyrobus. Power from the pickup arm drives the motor generator to spin the flywheel.

same pressure vessel previously described; in hydraulic systems where oil is the working fluid a receiver is provided with a flexible diaphragm separating the oil from a trapped volume of air or gas. As oil is forced into the receiver the diaphragm is pressed back and the air on the other side compressed; when the oil flows out again the energy stored in the air is recovered.

Compressed air can also be used for the storage of electrical energy. In one system at Huntorf, West Germany, off-peak electric power is used to pump air into an underground storage cavern. It is then stored under pressure until required to drive two 290 MW turbogenerators which run for two hours a day to meet peak power requirements. As well as using natural or artificially created underground caverns, investigations are being made into the possibilities of using the natural storage left when a gas well has been exhausted. Alternately reinforced concrete storage vessels could also be constructed to contain the pressurized air.

A recent development is the use of very high pressure pneumatic storage batteries. These use containers machined from solid metal to hold gas at very high pressures. Air is used for pressures of up to 600 million lb/sq in. (415×10^5 bar), a helium and air mix for pressures up to 10 billion lb/sq in. (6896×10^5 bar) and pure helium up to 17 billion lb/sq in. ($11,725 \times 10^5$ bar). The stored power can be released immediately or over a period of time and is safe for use where electric sparks would be dangerous. At present these pneumatic batteries are mainly used for military applications, such as driving gyroscopes in guidance systems.

Electric storage batteries

The most common type of storage battery is the lead-acid ACCUMULATOR which has to be charged up over a period of about 10 hours; when discharged about 90 per cent of the actual current storing capacity (amps × time) is recovered but, as the discharge voltage is lower than the charging voltage, the actual energy recovered is only about 75 per cent of that used to charge the battery.

For vehicle uses the weight of the battery is particularly important and great efforts have been made to reduce the weight for a given output. This aim has partly been met by the development of alkali batteries, which have nickel and cadium or nickel and iron plates in a potassium hydroxide solution. These batteries are very rugged both mechanically and electrically and have found considerable application to electric vehicle drives, though they are still heavier than the equivalent lead-acid batteries. The current recovery is about 75 to 80 per cent but the energy return is only 60 to 65 per cent.

High power densities and cycle efficiencies are offered by high-temperature batteries such as the lithium alloy-iron sulfide and sodium-sulfur batteries. The sodium-sulfur design has a working temperature of over 572° F (300° C) and has sodium and sulfur electrodes, which are in a liquid state at the working temperature, and an alumina electrolyte which is solid. The output per unit weight (around 140 Wh/kg) is more than five times that of the established lead-acid battery.

Another form of electrical energy storage involves the electrolysis of water to form hydrogen and oxygen gases which are stored and then allowed to recombine in a fuel cell to generate electricity. Such fuel cells are used to supply power on the Space Shuttle.

The increasing use of portable electronic devices, such as radios and calculators, has led to a requirement for small types of rechargeable battery. Sealed nickel-cadmium cells are often used but higher energy densities are offered by zinc-silver oxide cells. An alternative approach to meeting small power requirements in portable equipment is to use large capacitors rated at several Farads. These have particular advantages when repeated charge/discharge cycles are involved.

Thermal storage

Electric heating is an inefficient process because the original generation of electricity involves conversion from fuel to heat, to mechanical power, and then to electric power with efficiencies of 40 per cent or lower. Conversion of the electric energy back to heat is clearly wasteful but the overall inefficiency can sometimes be justified on account of cleanliness and convenience. This is particularly so if the efficiency of the overall system can be increased by using power at the low demand periods. Such systems can be used for heating by means of storage heater systems.

Electric resistance wires (similar to those found in an ordinary electric heater) are arranged to pass through a large mass of refractory (heat resisting) material which has a high heat capacity (absorbs a large amount of heat for a comparatively small temperature change). Air passages are provided through the refractory material and the whole assembly is contained in a casing which is thermally insulated so that heat will not be lost to the surroundings. Arrangements are made to allow air to flow through the whole assembly as required and the air then carries away heat to the surrounding space. The flow of air can be controlled by vanes which are generally operated automatically according to the outside temperature. Such systems are normally operated with a time-controlled switch so that heating only takes place during the night but heat may be released at any time. The efficiency of all electric space heating systems is in fact 100 per cent as there is no loss of heat except to the surrounding space where the heat is required.

Large capacity thermal storage at practically constant temperature can be achieved if, instead of the blocks of refractory material, containers of mixtures of suitable salts are used. The salts are chosen so that the mixture has a melting point at the required temperature. Where heat is available for storage, it is absorbed to melt the salts. It can later be released by allowing the salts to resolidify. This

Below: In Britain, electric storage radiators absorb heat in the night to give it out during the day.

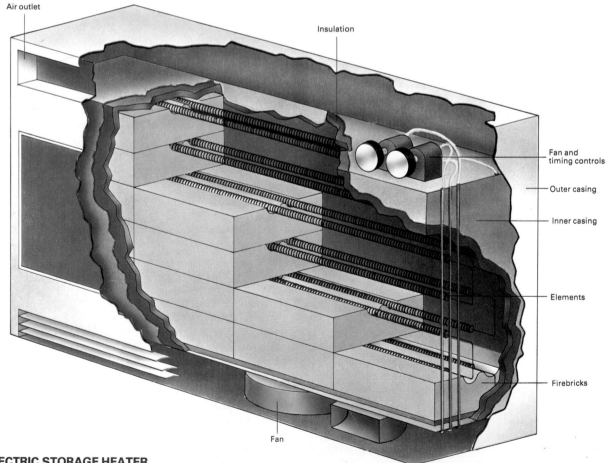

Air outlet

Insulation

Fan and timing controls

Outer casing

Inner casing

Elements

Firebricks

Fan

ELECTRIC STORAGE HEATER

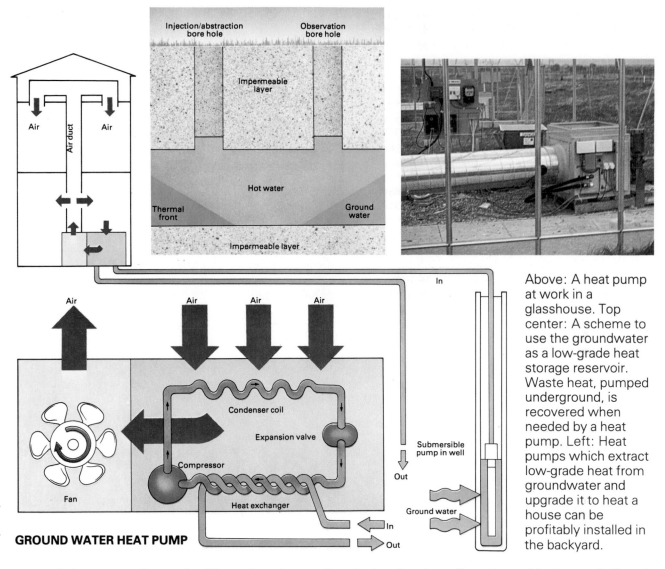

Injection/abstraction bore hole

Observation bore hole

Impermeable layer

Hot water

Thermal front

Ground water

Impermeable layer

Air

Air

Air duct

Air

Air

Air

Air

Air

Condenser coil

Expansion valve

Compressor

Fan

Heat exchanger

In

Out

In

Out

Submersible pump in well

Ground water

GROUND WATER HEAT PUMP

Above: A heat pump at work in a glasshouse. Top center: A scheme to use the groundwater as a low-grade heat storage reservoir. Waste heat, pumped underground, is recovered when needed by a heat pump. Left: Heat pumps which extract low-grade heat from groundwater and upgrade it to heat a house can be profitably installed in the backyard.

type of storage can be used with various types of heat source including solar radiation, but has not yet been used extensively.

There is also a thermal equivalent of the electric battery, which uses a sealed sachet of stabilized SUPERSATURATED SOLUTION of Thermogel (sodium acetate trihydrate) which can be caused to crystallize when heat is required. The sachet is initially at ambient temperature but as the solution solidifies the latent heat of crystallization is given up in a controlled manner. This crystallization is a reversible process so heat can be stored by reheating the sachet to remelt the Thermogel.

Mechanical energy storage
There are a number of systems for storing energy by purely mechanical means though they are rarely used on a large scale. One of the simplest forms is to arrange for railway tracks to slope as they enter and leave each station. In this way the kinetic energy of

a train when traveling at speed between stations is converted into potential energy as it climbs the slope (and therefore slows down) when entering the station. On leaving the station it runs down the slope on the other side; potential energy is reconverted to kinetic energy, helping the train accelerate.

There have been many other schemes intended to achieve a similar result by storing surplus energy in rotation of a flywheel (as kinetic energy) during periods when brakes might otherwise be applied, and recovering the energy when required.

Early hunters used a form of strain energy storage in bows and arrows, catapults and many other similar devices. The original source of mechanical energy is the hunter's muscle power, which is stored, with 100 per cent efficiency, in the springiness of the wood or the catapult elastic.

See also: Battery; Energy; Energy sources; Flywheel; Fuel cell; Hydroelectric power.

Engraving and etching

Engraving and etching on metal are old crafts. Although their use as printmaking processes dates only from the fifteenth century, they were in use long before that by metal workers as a means of decorating arms, armor and jewelry by incising a design into the surface of the metal.

In the middle of the fifteenth century the idea of printing from engraved or etched plates occurred at almost the same time in various places in Germany and Italy. In order to keep a record of their original designs, goldsmiths filled the engraved or etched lines in a metal plate with a black greasy mixture and pressed a piece of paper against the plate to pick up the ink from the grooves.

With the invention of photography, the crafts of engraving and etching as a means of reproduction of pictures died out, but these processes are widely used by artists as media in their own right. The techniques are defined under the general heading of *intaglio* processes. Plates of steel, copper or zinc less than ⅛ in. thick are used. Grooves, pitted areas and textures are *bitten* into the plate with acid (*etching, aquatint, soft ground*) or directly cut or scratched by the artist using sharp tools (*engraving, drypoint,*

mezzotint). To obtain a print from such a plate, ink is pushed into the grooves and the surface of the plate is wiped clean with muslin. The plate is placed on the bed of a copperplate printing press; a sheet of dampened paper is laid on the plate and backed by several layers of fine felt blankets. When taken through the press under pressure, the blankets force the paper into the grooves, where it picks up the ink. Despite all the progress in commercial printing presses, no one has managed to invent a satisfactory automatic process.

Engraving

The tool used is called a *burin* or *graver*, and its action is limited to lines and dots. The art of the engraver has been to manipulate this limitation to advantage, building up a variety of tones and textures by varying the width, depth and size of the lines and dots. An engraving is differentiated from other types of prints by its sharpness and clarity.

Below: In line engraving, a design is made by cutting directly into the surface of a copper plate. In etching, the design is scored through a wax coat and treated with sulfuric acid which eats away at any exposed copper to leave the required pattern. In both cases the patterns are printed similarly.

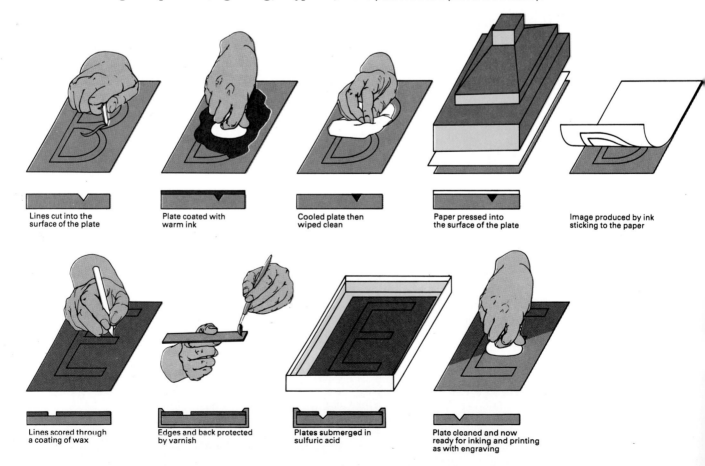

Lines cut into the
surface of the plate

Plate coated with
warm ink

Cooled plate then
wiped clean

Paper pressed into
the surface of the plate

Image produced by ink
sticking to the paper

Lines scored through
a coating of wax

Edges and back protected
by varnish

Plates submerged in
sulfuric acid

Plate cleaned and now
ready for inking and printing
as with engraving

The tool is a short, highly tempered steel bar about ¼ in. thick, square or lozenge-shaped in section. One end is bent up into a mushroom-shaped handle, so that when the tool lies flat on the plate the handle comes up comfortably into the palm of the hand. A facet is cut at a 45 to 60° angle on the other end, and the engraving is done with the sharpened tip of the tool by holding it at a shallow angle to the plate and pushing it slowly forward. As the tool cuts grooves in the metal it generates a chip which is removed at the end of the line with the side of the tool. The bigger the angle of the tool to the plate, the wider and deeper the incision. The greater the amount of ink which the line will hold, the darker it will print. Curved lines are executed by turning the plate rather than the tool. Round dots are executed by placing the tip of the tool on the plate and turning the plate in a complete circle; it is also possible to execute triangular dots by pushing the tip of the tool into the plate and pulling it out.

Below left: Using a burin to engrave a plate. Below right: A demonstration of the technique using a rocker. Bottom left: The artist has inked the plate and is now wiping it off. Bottom right: The plate on the press with paper on top. The press forces the paper into the inked lines.

Mezzotint

Mezzotint is a tonal technique, invented in the middle of the seventeenth century and used to copy paintings. It is a time-consuming process, requiring elaborate preparation of the plate. The plate is prepared with a tool called a *rocker*, made of hard steel and chisel shaped. The end of the tool is rounded, and has grooves in it from 45 to 120 to the inch; one side of it is sharpened so that the sharp edge is serrated by the grooves. This tool is then rocked across the surface of the plate in as many as 80 different directions until the entire surface of the plate is covered with a texture and so shiny dots of polished surface remain. When the plate is ready, other sharp tools called burnishers and scrapers are used to scrape away the texture to varying depths. The ink held by the plate determines how dark it will print: the more the texture is scraped away, the lighter that area will print.

Etching

The principle of etching is to protect the surface of the plate with a thin layer of acid-resistant substance called an etching *ground*. The plate is heated and this ground, made of asphaltum, beeswax and resin, is melted onto the surface of the plate and evened out with a roller. When the plate is cooled it

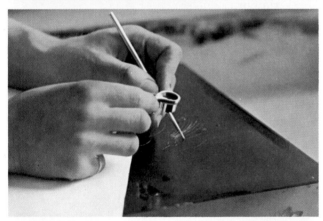

is smoked with tapers to harden the ground and to make it evenly black in order to facilitate seeing the drawing. The drawing is scratched through the ground and the plate is bathed in an acid solution. The longer the plate is left in the solution, the deeper the lines, and the darker they will print.

Aquatint

Aquatint is a tonal etching method. It was invented in the eighteenth century in France. The surface of the plate is covered with a thin, even layer of fine powdered resin. The plate is heated from beneath and as the powder melts it forms a uniform surface of acid-resistant granules which firmly adhere to it. The acid will bite into the plate around each granule; the deeper the bite the darker the tone will be. For different tones, the plate is removed from the acid and the lightest tones stopped out with liquid varnish on a brush. This process is repeated until the varying tones are achieved. The resin is washed off the finished etching with methylated spirits.

Sugar aquatint

In this technique the drawing is made with a brush using a solution of sugar and India ink. The plate is then coated with diluted acid-resistant varnish which, because it is diluted with turpentine, does not disturb the sugar in the drawing. The plate is then immersed in water; the sugar swells and dissolves, lifting off the varnish in the drawn lines and leaving the image exposed. An aquatint ground is laid and the plate is bitten in the acid.

Other applications

The principles of engraving and etching are used in many applications by science and industry. Currency is usually printed by engraved plates to make it hard to counterfeit. Toolmakers use etching and engraving methods to incise trademarks, measurements and other vital data on their products. Metallurgists etch metal samples directly without using a resist (ground) before examining them under microscopes to make their grain characteristics stand out better. Photo engraving (or gravure) processes are used for the reproduction of photographs by a printing press.

See also: Armor; Counterfeiting and forgery; Ink; Newspaper production; Printing.

Top left: Engraving a design on a silver plate in Indonesia. Second from top: The etching process. A plate is heated to harden the etching ground which creates a uniform black work surface. Third from top: Using precision tools and a magnifying glass, the artist scratches the design into the black ground. Bottom: Acid is pushed carefully into the lines with a feather.

Enzyme

Enzymes are biological catalysts which are vital to living organisms. Without enzymes, the process of turning the food an organism eats into energy that can be used to keep its cell functioning, for example, would be impossibly slow. It is not known how many enzymes exist; more than 1000 have been listed and many more doubtless await discovery.

Those enzymes that have been isolated and identified in a pure state have proved to be proteins and although most enzymes have not been positively identified as proteins, those conditions and substances that denature, or precipitate, proteins also inactivate enzymes. Some of these conditions and substances are extremes of acidity or alkalinity, the salts of heavy metals, ultraviolet light and high temperatures. Most catalytic reactions speed up when heat is applied, and enzyme-assisted reactions are no exception. If the cell is heated, the reaction speeds up until the optimum temperature is reached. If the cell gets any hotter, the enzymes become progressively inactive and the reaction slows until eventually, about 120° F (50° C), enzymes are completely denatured and unable to catalyze any reactions. The optimum temperature in humans is between 86° and 104° F (30° and 40° C). Within this range, there is maximum reaction speed with no damage to the enzymes. The destruction of enzymes at high temperatures is probably one reason that a prolonged high internal temperature, as in a high fever, can prove fatal.

Metabolism

Enzymes promote the chemical processes collectively called metabolism. There are two aspects to metabolism: anabolism, which is the synthesis of complex molecules from simple ones, and katabolism, which is the degradation or breaking down of large molecules into simpler ones. The enzymes concerned in these processes may be classified in the following way: *transferases* transfer a substance from one molecule to another; *isomerases* rearrange molecules; *lyases* split complex substances into simpler ones; *hydrolases* act like lyases but use water in the reaction; *ligases* or *synthetases* are concerned with the synthesis of substances; and *oxireductases* are concerned with oxidizing and reducing substances.

Enzymes, unlike other catalysts, are specific — each is concerned with only one reaction and acts only upon a *substrate*. For example, the enzyme maltase is involved only in converting its substrate maltose into glucose. There are a few which act on more than one substrate but even these take part in a limited range of reactions. The actions and interactions of enzymes are complex and the absence of just one enzyme in the body can have catastrophic results for health.

There are many instances in which many enzymes are required in balance to maintain health. Any imbalance either because of overproduction or deficiency will upset the metabolism and can cause a chain of disorders.

See also: **Cell; Digestive system; Metabolism.**

Right: Yeast enzymes, when fermented, can produce bacterial protein as an end result. Here, yeasts grown on methanol (produced from natural gas) pass through an oxygen-rich fermenter vessel. After drying, a protein-rich animal food results.

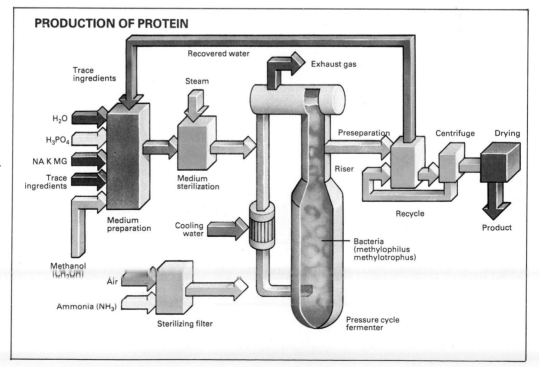

PRODUCTION OF PROTEIN

Trace ingredients

Recovered water

Exhaust gas

Steam

H_2O

H_3PO_4

NA K MG

Trace ingredients

Medium sterilization

Medium preparation

Cooling water

Methanol (CH_3OH)

Air

Ammonia (NH_3)

Sterilizing filter

Preseparation

Riser

Recycle

Centrifuge

Drying

Product

Bacteria (methylophilus methylotrophus)

Pressure cycle fermenter

Ergonomic design

Any realistic design requires that the designer makes tradeoff decisions. People who buy goods expect them to be reliable, efficient, safe, easy to use and value for money. The manufacturer would probably add that they should be eye-catching, easy and cheap to manufacture, and appeal to a large section of the population. The difference between a good design and a poor one is the nature of the trade-off decisions, and these in turn depend on the criteria the designer sets before starting to work. For example, a designer may decide that low-cost manufacture is the most important thing, and may sacrifice some aspects of safety and reliability or other factors to achieve this aim.

Ergonomics – the principle that underlines ergonomic design – is concerned with the interaction between humans and the environment in which they live and work. It is concerned with the home, offices, factories, hospitals and schools, and the vehicles that transport people between them. Its goals are to insure that these interactions occur with ease, in safety, and without error, to the benefit of the individual and society. If they can be made pleasurable, so much the better.

The interactions ergonomics considers are those directly between a user and an object, bearing in mind the range of shapes, sizes, knowledge and intentions of the user population, and also whether the interaction is intentional or accidental. It also considers the effect of this interaction on the society in which it is embedded. Ergonomic design is the process by which the designed environment is matched to the characteristics of people, and draws upon a wide body of knowledge and methods to achieve this.

Ergonomics insures that automobile seats, for example, can be adjusted so they are comfortable for people of all sizes and shapes, and an ergonomically designed can-opener will not injure the user's fingers. Ergonomic design goes even further; it includes not only everyday consumer goods, but also complex systems, such as the cockpit and life-support systems of the Space Shuttle, and the design of control rooms for nuclear plants. The philosophy behind ergonomic design is that things should be safe, easy and comfortable to use; a person should not have to battle with the tools to do a task. Its importance lies in the fact that all technological developments are intended to benefit people.

Right: This modern car is ergonomically designed to allow drivers large or small, male or female, to reach the wheel, pedals and controls easily and without effort by adjusting the seat or wheel.
Below: Badly placed controls on this tractor force the driver to bend over to reach them, causing unnecessary stress and fatigue.

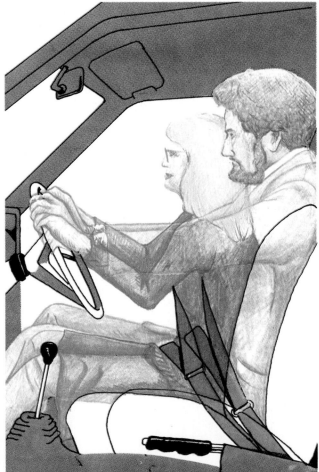

Ergonomics is not a new practice. When the early cave dwellers took to chipping flints to make hand tools, they were practicing ergonomics. Ergonomics embraces knowledge from a number of sources. The word itself was coined in 1948. Derived from the Greek, it means combining work with natural laws. It is also known as Human Factors, and, occasionally, Engineering Psychology, and it draws upon physics, physiology and anatomy, psychology, design, statistics, computer science, and engineering as its main sources.

Throughout the ages, the commonsense approach practiced by cave dwellers has continued. Most hand tools, especially those used by craftspeople, have evolved by trial and error to be most effective, albeit requiring a little skill in their application. During this century, however, technology has moved to the point where native wit, commonsense, and trial and error are no longer enough, and must be supplemented by more formal, scientific knowledge and approaches. Today, manufacturers design objects that will be used by millions of people, among whom there is a great diversity of physical and mental abilities. Without careful thought and appropriate

Below: A designer's sketch, showing one idea for a solid state instrument panel. The real breakthrough is in devices such as trip computers and head-up displays, projecting information onto the windshield so that the driver's line of vision need not be shifted.

methods, an object might be of limited use in some hands or might injure the user if it is not used in the way intended by the maker.

To design the interior of an automobile, for example, there are several ergonomic considerations, besides the other important factors, such as esthetic appeal and cost. The people who use the automobiles are of prime importance – the designer must know about their size, both when sitting and when getting in or out. Equally important is the amount of space they need to feel comfortable and, for drivers, easy access to the pedals on the floor and the controls on the dashboard and steering column. The driver must be in the correct position, whatever his or her size, to have good vision in all directions. Then the seat must be comfortable for a long journey; it should not transmit much vibration and should hold the driver in place around corners.

There is also the problem of displaying information about the automobile to the driver, who should be able to see or hear important information about errors, and without having to look down inside the car for too long. This requires that the displays, such as the odometer and other instruments, are easy to understand and are clearly visible. In short, the designer must understand the functions of vision and hearing, especially when the driver might be tired. Together with the many other considerations, these factors indicate that arriving at a design suitable for most people to use is not simple.

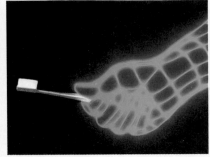
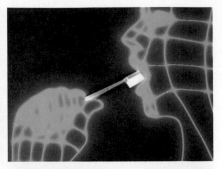

Above, left to right: Solid modeling can even help in the design of a toothbrush. The computer can be used to design the arrangement of the bristles and see how the brush is held in the hand – important for correct brushing action. The computer can also show how the brush will cope with the various shapes of teeth.

Computerization

Modern designers rely extensively on the use of computers around which an entirely new discipline has developed – Computer-Aided Design (CAD). This is a design station that has facilities for three-dimensional modeling of an entire system, including user and machine. The model enables the designer to originate and check designs quickly, before building prototypes. Some of the problems are to provide a system that meets the needs of the designer. For example, there should be provision of an easy dialogue for the designer to talk to the computer, and facilities to change details of the design. More importantly, care must be taken to insure that the computer graphics do not mislead the designer. For that matter, what does the designer really need from the system, and how easily can designers be trained to use the system once it has been built? In this case, ergonomic design is concerned with more abstract matters, many of which occur inside the designer's head.

Some companies that have installed CAD systems have had to change their management structure as a result (a few had planned the appropriate changes). Once people have been provided with different tools to perform a task, the nature of the task is changed. In design, one of the outcomes has been that the designer has taken on a more important role in the design office, and chief designers have had some of their authority reduced. Because of the greater capacity of the design system, the relationship between design and production can also change. Consideration of such changes is a part of ergonomic design, too, for there is little point in a company introducing a radically new tool if the resulting upheaval is to the company's detriment in

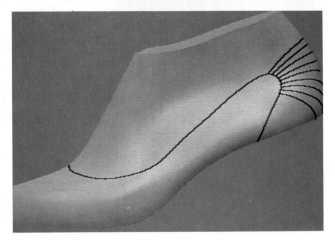

the long run. As management consultants have said, one of the quickest ways to go bankrupt is to introduce computer technology without adequate planning.

The methods used by ergonomists include scientific experiments, on-site user trials, interviews and questionnaires, expert judgments and performance logging. Traditional design methods are also used to establish and evaluate design constraints, and to develop and evaluate design solutions. In contrast with usual design practice, ergonomic design places great emphasis on user trials and user evaluations, for these are what prove the design in normal use and show up the design errors.

The value of ergonomic design is not always easy to quantify, at least in cost accounting terms. Clearly, however, in the high-technology world, ergonomic design is a positive advantage. Today it is rare for any piece of military equipment to be accepted into any of the armed forces in the West

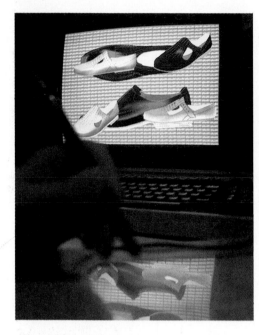

Left and above: With CAD, the designer can draw the foot and shoe shape, color in the leather and test the effects of walking.

Above: Using a light pen to adjust a computer model of a tractor. Right: Computers help design spacesuits.

without a full ergonomics evaluation. It is also becoming rare in the Information Technology world to find an advertisement that does not make great play of words such as user-friendly, usability, and ergonomics. This, in part, reflects the recognition by manufacturers that as software and hardware become more versatile, what will distinguish one manufacturer's product from another will be its interface with humans – not what is inside the box.

Occasionally, there are evaluations. For example, a large hotel chain in the U.S. estimated that, by good design of documents and attention to data input, they could reduce by a second the usual 150 seconds to make a hotel reservation, saving $24,000,000 per annum. But it is in the causes of accidents that the real value of ergonomic design is seen. On one space mission, owing to an error in keying data into a computer, the Space Shuttle dropped its fuel tanks 7500 miles (12,000 km) from the intended point.

On a more serious note, the crash of a New Zealand DC-10 aircraft into Mt Erebus in Antarctica and the deviation of the Korean Airlines Boeing 747 that led to it being shot down by the Soviet air force have been attributed to errors in setting navigational information into computers. It should be noted that flying is by far the safest means of transportation. Nevertheless, the consequences of small errors are horrific. It is by good, careful ergonomic design that such errors will be reduced.

See also: Assembly line; Automobile; Car safety.

Erosion

Erosion is the process of displacement of soil materials from the surface of the Earth. It results principally from the passage of wind, water or ice over or against the land surface. Erosion by ice occurs in sparsely populated areas, and is of little economic or social consequence. Erosion by wind and water is widespread, and in fact is exacerbated by modification of the natural environment for economic exploitation. An indication of the scale of the problem is given by figures from the U.S. Soil Conservation Service which show that in the period 1977–82 some 1.2 billion tons of soil were lost in the U.S. each year. It is estimated that over 40 per cent of U.S. farmland is losing soil faster than it is replaced by the normal organic processes. To a large extent, this problem has been hidden by the fact that the considerable increases in crop yields due to the use of fertilizers far outweigh the loss in production due to erosion.

Land surfaces may be subject to erosion by wind and rainfall, depending on factors such as climate and vegetation. Water erosion may be restricted to river channels or coastlines where the water and the land abut.

Control of soil erosion

Erosion by wind, known as deflation, is confined to areas of relatively flat land where rainfall is sparse, or to flat areas which are heavily cultivated. In such areas, winds are strong and frequent, and there is

Above: Wind erosion constantly shifts the sands of the desert. Left: A Brazilian forest destroyed by fire is replanted to create windbreaks.

not enough water in the soil to cause a cohesive effect. The speed of the wind can be reduced by *windbreaks*. A windbreak will reduce the speed of the wind at ground level for a distance of about five times its height on the windward side and twenty to thirty times on the leeward. The windbreak must, however, have a certain permeability to the wind, or serious turbulence (eddy currents) may result around the ends of the windbreak, which will result in local erosion.

Windbreaks usually consist of rows of trees planted at right angles to the prevailing winds. For maximum effectiveness the rows should be no more than about 1650 ft (500 m) apart. Natural windbreaks have certain drawbacks: they compete with crops for moisture, nutrients and space, they may harbor pests, and at the time of the year when crops are not growing the foliage on the trees may not be dense enough to protect the soil from the wind. Sometimes scientifically designed artificial windbreaks of wood or other materials are used.

Above: Erosion by ice and melting snow can play havoc with mountaintops; resulting surface disturbances can lead to avalanches and flooding.

The surface of the ground is best protected from erosion by vegetation. This reduces wind speed at ground level and the roots help to bind together the particles of soil; the absence of vegetation is the reason cultivated soil is particularly subject to erosion during the soil preparation and sowing season. Recent methods of sowing crops allow sowing to be done directly into the stubble and waste plant matter left from the last growing season, a technique known as *stubble mulching*. *Strip farming*, a method of cultivating alternate strips 300 to 600 ft (90 to 180 m) wide, not only prevents wearing out the soil of nutrients by overfarming it, but also helps prevent erosion, because the planted strips act as windbreaks. It is also possible to protect the soil by spraying it with bitumen emulsion; this is an especially useful technique in areas where a very sparse rainfall restricts the growth of a continuous vegetation cover.

Shifting cultivation, where the land is left fallow for several years to allow poor soils to recover, helps

to minimize erosion. However, population pressures are reducing the amount of time for which the land is left fallow, with a consequent increase in erosion. The widespread felling of trees, for commercial exploitation of the wood or for use as fuel, is another serious cause of erosion.

The intensity of soil erosion by water depends on the velocity as well as the volume of water flow over the ground, and therefore is worst in areas of steep slopes and heavy rainfall. Here again, stubble mulching is useful, because it reduces rain-splash erosion, which results from the impact of raindrops on the ground, and also retards surface water flow. The use of *cover crops* is also widespread. In the tropics, this may be a perennial tree beneath which other crops are grown; in more temperate climates, a cover crop will be a winter annual which protects the bare ground when crops are not growing.

On slopes, *contour plowing* is practiced. Plowing along the contour rather than up and down the slope reduces the total run-off because water is stored in the furrow rather than channeled by it. (Chaneling of run-off down a furrow results in a rapid erosion causing *gullies*.) Soil erosion can be reduced more than 50 per cent by contour plowing.

Terracing of hillsides is another contour technique. A system of ditches and embankments is

Below: Contour plowing is used in agriculture to minimize erosion by collecting run-off in furrows. It maximizes land efficiency by preventing unplantable wet areas at a slope's bottom.

is past. However, heavy erosion can lead to the dams silting up, which considerably reduces their efficiency as flow regulators. Where erosion by running water is the problem, a scheme of combined soil erosion control and river control is used.

Where erosion of stream banks is a problem, walls or embankments are constructed to replace the vulnerable natural bank. The erosion problem is usually concentrated on the outside bank of a bend in the river, where protection is most often needed.

Coastal erosion
Coastal erosion can be spectacular, and the cost of controlling it very high. Fortunately only restricted areas of coastline need protection, usually sites of urban development where soft rock is exposed to erosive attack.

In many areas natural protection is afforded by beaches of sand or gravel; the slope of a beach dissipates the energy of waves. Beaches, however, are extremely unstable, and the material shifts alongshore. *Groynes* are constructed to impede this movement and to encourage the buildup of the beach. Groynes are barriers built out from the high tide mark, usually at right angles to the shore. They interfere with the alongshore drift, each groyne collecting material on its updrift side. *Boxed groynes* are a more unusual attempt to build up material. These are a network of shallow, closely spaced groynes with some elements parallel and others at

Erosion is combated by terracing hillsides (above) or interlocking spoked cement shapes (right) on seashores to absorb and break the energy of waves.

constructed along the contour so as to direct run-off on a low gradient path to the edge of fields where it is drained off into prepared channels. Terracing is an ancient agricultural practice found in hilly areas of Central and South America, and southern and eastern Asia, and is being adopted on a large scale by other countries.

River erosion
Large-scale erosive activity of rivers and streams is likely to be restricted to periods of peak flow following heavy rainfall or melting of ice and snow at higher altitudes. Control aims to reduce the intensity of run-off into the stream system, and is often the same thing as flood control. Measures controlling soil erosion on the land near the river banks help to control peak flows; in addition, *afforestation* is frequently carried out. A woodland cover, especially near the headwater sections of a river, interrupts much of the run-off and lowers the level at the peak. Water flow is controlled by *dams*, which regulate the flow by filling up during peak periods; the water can be slowly discharged when the peak

right angles to the shoreline, forming a series of box sections. Beach material is thrown into the boxes by waves but the boxes inhibit movement of it during destructive storms. Linking groynes to sills out at sea encourages naturally formed beaches; sand is deposited by waves breaking over the sills, but the water can flow back out.

On some coastal areas of North and South America, a new development of groynes in the form of self-sealing plastic bags has been successfully installed. The bags are designed to be of a given porosity according to local requirements; the waves deposit sand on the installation and the water filters through. Pyramid groynes built of these bags are often more successful than vertical ones.

Sea walls or banks are an additional method of coastal defense. Walls which gently slope break the force of incoming waves and do not have to withstand the full force of storm waves. In resort areas, walls with nearly vertical faces are sometimes built because they take up less space; these are sometimes designed with a "nose" at the top to throw the wave back instead of allowing it to splash over. The wall lasts longer if combined with groynes.

• FACT FILE •

- Salt crystals in some limestones and in building mortars can cause erosion. Egyptian temples at Luxor and Karnak have been severely damaged by salt seeping into crevices and recrystallizing repeatedly over the centuries. In Antarctica, boulders have been pitted with holes up to 6 ft (2 m) in diameter as salt crystals have grown, prizing minerals away from the structure.

- Surveys carried out in the U.S. in the wake of the droughts and dust storms of 1933–34 showed that almost 75 per cent of a total arable land area of 414 million acres (168 million hectares) were seriously damaged by wind erosion. Some 200 million acres (80 million hectares) had lost half or more of the topsoil.

- The building of shoulder-to-shoulder hotels along Miami Beach, and the sea walls and breakwaters to protect them, caused localized wave action which stripped the beach of sand for 20 years. Artificial replenishment in the 1970s cost many millions of dollars, and will have to be repeated regularly every ten years or so to avoid further wave erosion.

Above: Tough grass provides a network of roots below and cover above to protect sand dunes.

The use of groynes or walls must be carefully engineered; erosion is often made worse at the ends of the protecting structures. Groynes especially, by preventing natural drift, may protect one section of beach but cause the next to be completely destroyed by depriving it of drifting material.

Beach erosion can also be significantly affected by other factors, such as the construction of dams on rivers. Such dams reduce the amount of material carried downstream and deposited along the coastline, so counteracting wave effects. Thus erosion of the Nile Delta has increased considerably since the building of the Aswan High Dam in Egypt.

All erosion can be defined as a natural, continuing process of land sculpture and development throughout geological time. Humans may wish to reduce it or control it in a given area, but they cannot hope to stop it. What they must be particularly aware of is the need to control erosion that they induce by their own modification of the landscape.

See also: **Agriculture technology; Farming, intensive; Fertilizer; Rain and rainfall; Soil research; Weather control.**

Escalator

The modern escalator provides the most efficient method of moving large numbers of people from one level to another at a controlled and even rate. In its simplest form it consists of a series of individual steps mounted between two endless chains which move upward or downward within a steel frame.

The pioneers of the escalator were Jesse Reno and Charles Seeberger, whose designs were produced independently in the early 1890s, and in 1900 escalators were installed in Paris and New York. The first modern type of escalator, however, did not appear until 1921, incorporating the best features of both the Reno and Seeberger designs, and today escalators are commonplace in railway stations, airports and department stores.

Top section

The top section houses the electrically powered driving machine and most of the controlling switchgear. The driving machine consists of an AC induction motor running at around 1000 revs/min and driving the escalator through a worm type reduction gear. The main brake is spring-loaded into its on position, being held off by a DC electromagnet to

Below: The escalator at Centre Georges Pompidou in Paris, France, is not only functional but also integral to the design of the building.

enable the escalator to run. This arrangement is thus fail-safe, as in the event of a power supply failure the electromagnet will be de-energized and the brake will be applied by the springs. A hand winding wheel and a manual brake release lever allow the escalator to be moved by hand if necessary.

The controlling unit includes rectifiers to provide the DC supply to the brake, a contactor to start the motor, and control relays linked to safety switches which will stop the machine in the event of an overload, drive chain breakage, or an obstruction to the steps or handrail. The controller is also linked to the key-operated starting switches and the emergency stop buttons, and contains a device to prevent an up-traveling escalator from reversing its direction of travel in the event of drive mechanism failure. To reduce wear and running costs some escalators are fitted with speed control devices that run the machine at half speed when no passengers are using it. Photoelectric sensors are fitted at each end to switch the escalator to full speed when a passenger steps on and to return it to half speed when all the passengers are cleared.

Center and bottom sections

The bottom section carries the step return idler sprockets (toothed wheels around which the chain turns), step chain safety switches and curved track

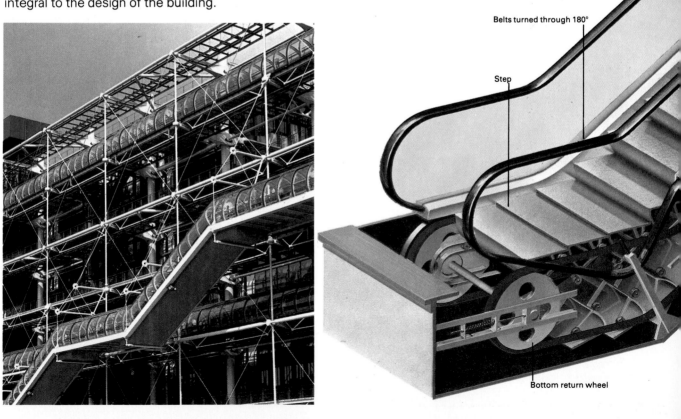

Belts turned through 180°

Step

Bottom return wheel

Below: A cutaway view of an escalator, which consists of a continuously moving series of individual steps mounted between two endless chains. The chains are driven around four toothed wheels by an induction motor. Each step moves on smaller wheels along rails which are positioned so that the steps fold to give a horizontal surface at the top and bottom. The handrail is a moving belt of rubber and canvas.

Handrail

Drive shaft and wheels

Endless chain

Handrail drive   Ratchet wheel

Inner rail

Returning steps

Top pair of wheels

Bottom pair of wheels

Outer rail

Variations on the escalator include travelators, continuous horizontal belts often found in airports to move people through terminals quickly, and the new curving escalator by Mitsubishi, designed with extra guide rollers to give a curved path.

sections to carry the motion of the escalator from the horizontal plane into the angle of climb, which is usually 30 to 35°. Between the top and bottom sections runs a welded box-type structure which carries the straight track sections.

The steps on which the passengers stand are normally assembled from aluminum pressure die castings and steel pressings mounted on a frame, usually of cast aluminum, which runs on rollers on

the main track sections and is driven by the two main chains, one on each side. Foothold on the step surface is provided by a cleated board faced with aluminum or rubber.

The moving handrails are made from layers of canvas covered with a rubber or plastic molding. They run in continuous loops in T-shape guides along the tops of the balustrades, at a speed closely linked with that of the steps. The balustrades and their skirtings allow a smooth passage for the steps, and all joints are securely masked. The running clearance of the steps has to comply with strict safety standards and the *combplates* (the metal teeth which project at the top and bottom of the fixed escalator base and provide the link between the moving treads and floor level) incorporate safety switches that will stop the escalator if any object is caught between the steps and combplate.

The tread width may be from 2 to 4 ft (0.6 to 1.2 m), and the speed of the operation from 90 to 180 ft/min (27 to 54 m/min). Running at 145 ft/min (44 m/min) a single escalator powered by a 100 hp motor can carry up to 10,000 passengers an hour. A variation of the escalator, the travelator, is a horizontal continuous rubber belt used for moving passengers around such places as airports and train stations.

See also: Electric motor; Photoelectric cell.

Evolution

Most scientists believe that all life on Earth is the product of, and participator in, a gradual and never-ending process of change they term evolution. The theory of evolution, although closely associated with the name of Charles Darwin, has itself *evolved* over the years, with many scientists contributing their ideas.

As early as the late 1760s an engineer named Smith, while supervising the excavation of a canal, noticed the fossilized remains of long-dead creatures. He also noticed that different fossils occurred in layers, or strata. After studying these fossils, Smith claimed that he could identify them simply by looking at the stratum from which they came.

Someone else with interest in fossils was the geologist Charles Lyell, who studied rock formations, particularly in relation to volcanoes. He came to the conclusion that the Earth was many millons of years old, much older than had previously been thought. Lyell's findings wer published in 1830, about the time Darwin was undertaking his famous

Below: The tree of life. All life derives from single-celled *Procaryotes*. In the Precambrian era plants and animals began developing along separate paths. Marine animals developed first; then bony fishes evolved into amphibians which in turn developed into mammals including humans. The gaps in this record, however, make it open to dispute.

voyage in the Beagle. Darwin realized that the changes in species he observed would take many millennia and Lyell's estimate of the antiquity of the Earth worked in well with his theories.

In 1809, Jean Baptiste de Lamark, the French naturalist, had propounded his theory of evolution. His ideas concerned the inheritance of acquired characteristics and he suggested that if a creature had a *besoin*, or need and desire for change, that desire caused the change and the change could, moreover, be passed on to its progeny. Darwin himself toyed with a similar idea, "use and disuse heredity," which suggested that the organs of a creature gave off particles that collected in the sex organs and then, passed on to the next generation, affected their organs' structure.

The survival theory

Apart from Lyell, one of the scientists who had most influence on Darwin was the nineteenth-century political economist, Thomas Malthus, who said in his book *An Essay on the Principles of Population* that nature produced far more offspring than would live to become adult and eventually reproduce. He noticed that there were thousands of seeds produced by, say, just one plant, or eggs by just two frogs, yet over a period of time the number of mature plants and frogs remained about the same. This idea formed the basis of Darwin's theories on natural selection. It was obvious that within a species individuals differ, and Darwin said that those individuals who were better at getting food and

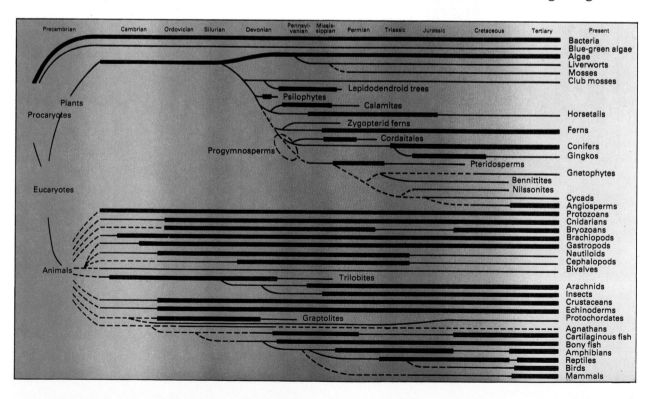

Right: The fossilized remains of the earliest known feathered bird, *Archaeopteryx*. This 150-million-year-old fossil, found in south Germany in 1861, only two years after Darwin proposed his theory of evolution, was crucial evidence in his support because it was a clear example of intermediates between one type of creature and another. Its reptilian tail, claws and teeth coexisted with birdlike feathers, a wishbone (which is a structure found only in birds) and chicken-like wings.

avoiding predators were more likely to survive and reproduce – this is the principle of The Survival of the Fittest.

At the same time Darwin was formulating his theory of natural selection, the naturalist Alfred Wallace had, as a result of his study of insects, reached much the same conclusion. He noted that although basically the same in structure, each variety of beetle differed in detail from the next according to its lifestyle.

In 1859 Darwin published his book *On the Origin of Species* in which he put forward his theory that all life on Earth had a common ancestry and that different species had evolved from that common ancestry to fill different ecological niches. Darwin suggested that one of the mechanisms involved in producing changes in individuals and eventually in species was sexual reproduction. This theory became established only in the light of the work of the Austrian Gregor Mendel, who worked out the Mendelian Laws of Inheritance, which defined the likelihood of a particular parental characteristic being transmitted to the offspring.

Inherited characteristics

Scientists have discovered that living cells contain paired structures called CHROMOSOMES, each containing many genes made up of deoxyribonucleic acid (DNA). These genes are inherited from the parents and are the blueprints of the organism.

Each parent contributes one chromosome to each pair. The genes along the chromsomes are matched to each other. The gene for eye color from one parent, for example, might be for blue eyes and the equivalent gene from the other parent might be for brown eyes. The brown-eye gene would dominate the recessive blue-eye gene, so the organism would have brown eyes.

When the organism with these two genes mates, only half the genetic material is passed on. This may be the half with the blue-eye gene or the half with the brown-eye gene. Depending on which genes come from the mate, the offspring could have either blue or brown eyes.

Occasionally, the genetic material undergoes a mutation – a radical change which produces a characteristic not present in either parent's genetic

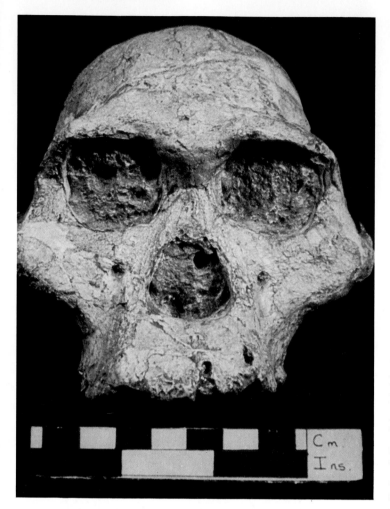

Above: Nearly two million years old, the massive skull and jaws of a robust australopithecine nicknamed "Nutcracker Man" were among the first to be accurately dated. Previously, the hominid was thought to be 600,000 years old.

makeup. Often, the organism does not survive, or, if it does, it is disadvantaged and does not live long.

Another change might be wholly advantageous, enabling the organism to compete more successfully for food and mates. One such mutation occurred in the European peppered moth, *Biston betularia*. This species lives on and around trees. It is thought that many centuries ago, the trees which were its habitat were dark-barked pines and the moth itself was dark and well-camouflaged. Over the years, deciduous trees with lighter-colored bark and a tendency to play host to lichens took over. The dark moth, now at a disadvantage, was easy prey but a mutant with peppered wings which blended better into the new habitat became the dominant species, until the Industrial Revolution. In those areas that became heavily industrialized, soot and fumes darkened the tree bark, killed the lichen and provided

again a habitat in which those moths that were darker could survive. In these areas, the peppered species is virtually extinct, the peppering being not only disadvantageous but also recessive.

Life on Earth

Many scientists believe that life arose on Earth from a primeval "soup" made up of such substances as water, hydrogen, methane, and ammonia. In a famous experiment, researchers passed electric discharges (to simulate lightning) through a similar mixture in the absence of oxygen – thousands of millions of years ago, the Earth had no atmosphere. They found that much more complex organic compounds were synthesized, including amino acids – the basic protein building blocks.

It is suggested that from such beginnings arose the most primitive forms of life – blue-green algae and bacteria. These unicellular creatures are alone in being prokaryotes – cells with no nucleus. The cells of all other organisms (eukaryotes) have a nucleus, which contains the genetic material. Eukaryotes have one other important difference from prokaryotes – sexual reproduction, giving the possibility of faster rates of change and increasing diversification. Blue-green algae, probably the first life on Earth, paved the way for other organisms by producing oxygen as a byproduct of PHOTOSYNTHESIS.

Geological periods

The time scale of evolution is almost incomprehensibly long. It is impossible to imagine time on the scale of hundreds of millions of years, so the history of the world is often compared with a year. On that scale, the first bacteria evolved around April and humans at 11:55 pm on December 31.

Another comparison would be to call one year an inch, then the 600 million years of the Earth's existence would be more than 9500 miles (15,200 km) and human's existence a mere 9 miles (15 km).

What is known of the history of life on Earth has been deduced from the study of geology and fossils, using advanced radioactive dating techniques. The Earth is thought to be about 4500 million years old. The first 3900 million years have been termed the Precambrian Era and it is generally agreed that life first appeared then, about 3500 million years ago.

Over the next 2000 million years, oxygen built up and an atmosphere was formed. Some of the oxygen in the higher layers of the atmosphere was turned into ozone, which helped to screen out harmful ultraviolet light, paving the way for life on land, away from the ocean. Very few fossils exist in rocks formed during this age, because the creatures were soft-bodied and may have simply decayed, but traces of unicellular and (toward the end of the period) simple multicellular organisms, such as worms and jellyfish, have been found.

The next era, the Paleozoic, has been divided into six main periods. The first, the Cambrian, has a rich and varied fossil record, mostly of invertebrates. Examples of these are the trilobites – bottom-dwelling marine creatures related to crustaceans. Also found in the Cambrian are sponges, worms and sea cucumber. Toward the end of this period, the first fish appeared.

During the Ordovician Period (525 million to 450 million years ago), invertebrate sea creatures became more complex. Together, the Ordovician and Cambrian Periods could be called the Age of the Higher Invertebrate, because they were the dominant forms of life. The first vertebrates appeared in this period.

The Silurian Period (450 million to 410 million years ago) is the start of the Age of Fish, a term more generally applied to the next period, the Devonian. During Silurian times, the first vertebrates, armored jawless fish, become common and it is thought the scorpion became the first air-breathing land dweller. Plants also moved onto the land around this time.

The Devonian Period saw the emergence of insects and amphibians. The amphibians are thought to have developed from the lobe fins – vertebrate fish with lungs and weight-bearing limbs which could leave the water for periods. The ancestry of the insects is still mysterious. This time also saw an increase in the number and variety of fish and probably the first sharks.

The fifth period, the Carboniferous, is divided into the Mississippian and Pennsylvanian Epochs, also known as the Lower and Upper Carboniferous Periods. During this time, dragonflies, the first winged insects, appeared and the Pennsylvanian was marked by the appearance of the first wholly land vertebrates – the reptiles.

Plant life was well developed by this time and the giant ferns, horsetails and others, covered vast

Above left: Anthropologist Mary Leakey shown with 3.75-million-year-old hominid footprints found at Laetoli in Tanzania in 1978. Made when the hominids walked over volcanic ash dampened by rain, the footprints are important evidence of bipedalism.
Above: Telltale cut marks, made by a stone tool in slicing meat from this antelope bone, indicate that it was once a hominid meal.

areas of swampy grounds. The decayed and fossilized remains of these plants, compressed and altered over millions of years, are today mined as coal. Peat is a product of the same process but of more modern times.

The sixth period of the Paleozoic Era, the Permian, marked the start of the domination of the Earth by reptiles, on land and in salt and fresh water. It also saw the extinction of the trilobite, and coniferous trees evolved.

The age of mammals

The Mesozoic (middle life) Era saw in its first part, the Triassic Period, the emergence of mammals. They are thought to have evolved from mammallike reptiles which arose in the Permian Era and appeared to die out around the time the true mammals appeared. The first dinosaurs, which were to dominate the next two periods, arose at this time.

In the Jurassic and Cretaceous Periods, the reptiles, particularly the dinosaurs, filled those ecological niches that later the mammals would make their own. They were of all sizes, from the size of a rooster to giant sauropods like Brachiosaurus, weighing about 50 tons and with an immensely long neck, in all about 60 ft (18 m) tall.

Probably the best-known dinosaur is *Tyrannosaurus rex*, the largest land-dwelling carnivore ever. It stood about 19 ft (6 m) high, had big teeth in powerful jaws and preyed on the herbivorous dinosaurs, such as the sauropods. Reptiles also ruled

The Galapagos Islands off the coast of South America, visited by the young Darwin in 1835, are populated by a vast number of endemic species – those found nowhere else in the world. Species such as this family of marine iguanas (left) and the ground finch (above) have developed in this "evolutionary hothouse."

the air in these times, including the pterodactyl and similar creatures with membranous wings.

The Jurassic Period saw the emergence of the first bird, *Archeopteryx*. The discovery of the fossilized remains of one of these creatures shortly after Darwin published *On the Origin of Species* was a great boost to his theory, because it was the first example found of a creature midway between two groups, in this case the reptiles and the birds. *Archeopteryx* had feathered wings and a reptilelike tail. It was not descended from the birdlike reptiles but from a common ancestor.

The Ichthyosaurs were probably the best-known of the marine reptiles, several fossils of which have been found. They swam like a fish, propelled by the tail fin. The plesiosaurs had four flippers and a long neck. It has been suggested that the alleged Loch Ness Monster sighted in Scotland is a member of this family that somehow escaped extinction with the rest of the dinosaurs at the end of the Cretaceous Period.

No one has satisfactorily explained the sudden disappearance of virtually all the large reptiles at this time (crocodiles alone survived). One plausible theory suggests a sudden climatic change which reduced the activity of these cold-blooded creatures, making them less efficient at feeding and also reducing the viability of their eggs.

About 60 million years ago came the dawn of the Cenozoic Era, which is divided into two periods – the Tertiary and Quaternary. These in turn are divided into seven epochs, from the Paleocene to the Holocene. These periods extend to the present day – the Age of the Mammals.

Many different species of mammals have evolved in these periods though some of these are extinct. The most successful mammal is the human. Primates (of which humans are one) appeared at the beginning of this period, in the Paleocene Epoch. Although the development of several mammals (for example, the horse) can be traced, the evolution of humans is not so conclusive. Apparently, the early primates were small creatures like tree shrews.

Traces of apelike animals have been found from the Oligocene, and fossils from the Miocene seem to link up to the modern chimpanzees and gorillas and the hominid *Ramapithecus* to humans. There is, however, a gap of about 12 million years before the Pleistocene. Early human-type creatures, *Australopithecus*, appeared about two million years ago, followed one million years later by *Homo erectus* (upright humans) with a larger brain and greater tool-using ability. Modern humans, *Homo sapiens*, appeared about 100,000 years ago.

There was presumably at least one transitional stage between erectus and sapiens but as yet it has not been identified. Up to about 15,000 years ago, humans had a huge skull, receding forehead and heavy eyebrow ridges. Later, these characteristics were less marked in *Cromagnon* humans.

See also: Amino acid; Brain; Cell; Earth; Genetic engineering; Geological techniques; Geophysics; Life; Oxygen; Reproduction, biological.

Explosive

The first pyrotechnic material to be used in war was Greek fire, invented by Kallenikos in AD 673. Although not strictly an explosive, being more of an incendiary substance somewhat similar to napalm, it was used with devastating success in several battles in the Mediterranean, and was a primitive but effective incendiary material probably based on petroleum oils and sulfur. Its secret was lost with the fall of the Byzantine Empire in AD 1453.

The origins of gunpowder (or black powder) are obscure, but the Chinese were probably aware of the properties of saltpeter (potassium nitrate) as early as the Chin Dynasty (221 to 207 BC), although they

did not develop its use beyond the firework stage until about the twelfth century. In the West this material was known as "Chinese snow," and the first recorded Western experimenter to establish the formula for gunpowder was an English monk, Roger Bacon. In AD 1245 he recorded it as a Latin anagram in his *De Secretis Operibus Artis et Naturae* (Secret Works of Art and Nature).

Its first use as a gun propellant followed in about 1320, and English troops used cannons against the French at Crécy in 1346. The formula has changed somewhat since then, and today the proportions used are 75 per cent saltpeter (potassium nitrate), 15 per cent charcoal (carbon) and 10 per cent sulfur. An alternative mix uses sodium nitrate instead of the potassium nitrate. It is not now used as a gun propellant as it burns too quickly, about 1312 ft (400 m) per second, its residue fouls the bore of the gun, and it produces too much smoke. On the other hand its combustion is too slow to produce the shattering effect of a high explosive. It is, however, widely used in fireworks and in blank cartridges, and it was used by the Russians in the retrorockets of the planetary surface probe sent to Mars.

Explosion and detonation

When an explosive substance is set off it undergoes rapid decomposition and releases large quantities of gas and heat. *Explosion* is a fast combustion, the burning layer spreading layer by layer through the material at the comparatively slow velocity of up to 1312 ft (400 m) per second, and although its rate increases with increasing pressure, it can be controlled. This reaction is often called *deflagration*.

In a *detonation* reaction there is an extremely rapid burning which produces a supersonic shock, or detonating wave, in the explosive substance. The detonation velocity is a characteristic of the explosive material itself and is unchanged by changes in pressure. It is usually between 6500 and 29,500 ft (2000 and 9000 m) per second. The detonation wave produces a very high pressure, about 700 tons per square inch (100,000 bar), and this exerts a severe shattering effect on anything in its path. The gases formed travel in the same direction as the detonation wave, so a low-pressure region is created behind it. Once the detonation wave has been started it cannot be stopped.

Explosives that react by deflagrating are called low explosives or propellants; they generate gases

Left: One of the most common uses of explosives is in quarrying. Skilled experts choose and position an adequate amount of explosives. Here the line of charges has been detonated to bring down a section of the face of a limestone quarry. The broken rock can then be easily excavated and removed by power shovels and high-payload trucks.

Above: Preparing an explosive charge. Holes are drilled into the surface to be blasted. A fuze is placed in each hole and connected by a length of wire. The explosive is funneled into the holes.

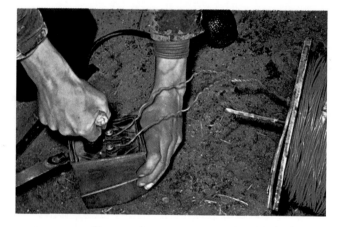

Above: After the holes have been filled, the trailing line is then wired to a detonator sited at a safe distance away. Electric impulses generated when its plunger is lowered detonate the charge.

Above: The specially designed fuze being ignited here is used by explosives experts working in separate locations on a blasting site to synchronize detonations. When the fuze burns to a given point, a red signal flare is released, giving others the go-ahead to detonate their charges.

at a slow enough rate for them to be used to propel rockets or shells. Explosives that react by detonation have the ability to shatter and are called high explosives. They are used in bombs and explosive shells, and in blasting operations such as quarrying and mining. Even higher explosive powers are obtained from the direct conversion of matter into energy by fusion or fission reactions, as in atomic bombs.

The essential properties of conventional chemical explosives are the velocity of burning or detonation, the explosion temperature, the sensitivity, and the power. For gunpowder, nitrate mixtures and nitro compounds, an absolute measurement is possible, but for the others it is usual to compare the explosive with a standard such as *picric acid*. Picric acid is taken to have a "value" of 100, with more sensitive or less powerful explosives having values lower than 100. The more sensitive explosives with values of around 20 are used as detonators to initiate explosives. Their impulses can set off the intermediary charges (those of moderate sensitivity, about 60) which in turn will initiate the reactions in the main charges, which are the least sensitive.

Modern explosives

The use of gunpowder as an explosive declined in the nineteenth century although it is still used for some applications – such as stone quarrying – where its relatively slow heaving action which splits the rock is more appropriate than the shattering effect of other explosives. Modern explosives are of three main types: those based on unstable molecules, such as the fulminates and azides; ammonium nitrate and the organic esters nitrocellulose, nitroglycerine and PETN; and the nitrocompounds, a large group which includes picric acid, TNT, tetryl and RDX.

Nitrate mixtures are now the most widely used type for blasting purposes, offering low cost and ease of handling, although the blasting power is less than for nitroglycerine or TNT. In many cases these mixtures are used as gels or slurries that consist of the nitrate mix, TNT (to improve the detonation characteristics), water and a gelling agent. The mix can simply be poured into the shotholes.

In 1845 Schönbein nitrated cotton with a mixture of nitric and sulfuric acids to give nitrocellulose $(C_6H_7O_2(NO_3)_3)_n$. Until 1875, when Sir Frederick Abel devised a method of pulping the cotton to give a more stable product, there were many accidental explosions associated with its manufacture. Its properties depend on the degree of nitration, which is hindered by the fibrous nature of the cotton, and it is usually a mixture of the di- and tri-nitrates. It is easily gelled by solvents and can then be pressed into the required shape, for example cord, flake or tube. It is very sensitive when dry (with a rating of

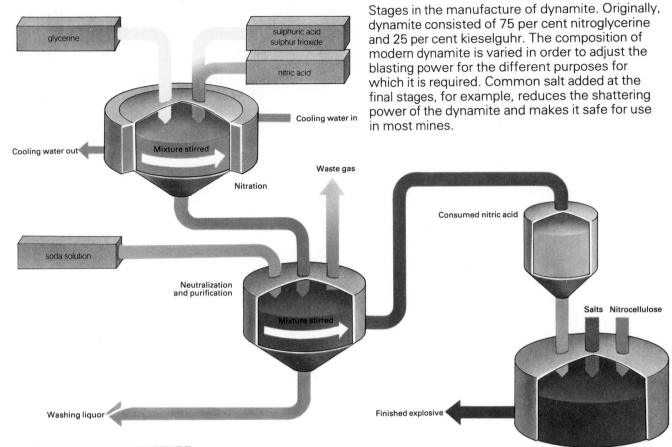

Stages in the manufacture of dynamite. Originally, dynamite consisted of 75 per cent nitroglycerine and 25 per cent kieselguhr. The composition of modern dynamite is varied in order to adjust the blasting power for the different purposes for which it is required. Common salt added at the final stages, for example, reduces the shattering power of the dynamite and makes it safe for use in most mines.

DYNAMITE MANUFACTURE

about 23), but less so when wet (about 120) yet can still be detonated. The velocity of detonation is 24,000 ft/s (7300 m/s) for dry nitrated cotton and 18,000 ft/s (5500 m/s) when wet.

The highly dangerous explosive liquid *nitroglycerine* (sensitivity 13) was first prepared in 1846 by Sobrero in Turin by nitrating glycerol with a mixture of nitric and sulfuric acids. In 1865 Alfred Nobel found that the liquid could be used safely if it was first absorbed into kieselguhr, a form of *diatomaceous* earth (formed by the fossil remains of a type of single-celled *algae*). He also succeeded in solidifying it by adding 8 per cent nitrocellulose to form a gel. He named these famous explosives dynamite and blasting gelatin. Nitroglycerine ($C_3H_5N_3O_9$) detonates at 25,400 ft/s (7750 m/s), has a power rating of 160, and an explosion temperature of 8000° F (4400° C).

PETN (*pentaerythritol tetranitrate*, $C_5H_8N_4O_{12}$) is a sensitive explosive (40) with a high power rating (166) and a detonation velocity of 26,500 ft/s (8100 m/s). It can be used as an intermediary charge but it is extensively used in detonating cords where its small critical diameter enables it to maintain a detonating impulse over a great distance, despite low filling densities.

Picric acid was first prepared by Woulfe in 1771, a hundred years before Sprengel demonstrated that it could be detonated by a mercury fulminate cap. Prepared by the nitration of phenol, it was the first of a large series of aromatic nitrocompounds to be discovered. As it is much less sensitive than nitrocellulose or nitroglycerine, but just as powerful, it safely withstands the shock of discharge from a gun, and in 1888 it replaced gunpowder as a shell filling. Apart from its value as a standard, picric acid, $C_6H_2(NO_2)_3OH$, is little used today.

TNT, *trinitrotoluene*, $C_7H_5(NO_2)_3$, is made by reacting toluene with a nitrating mixture of nitric and sulfuric acids. Its sensitivity is 110, power rating 95, and its velocity of detonation 23,000 ft/s (7000 m/s). It is cheap and easy to make and widely used to fill shells and bombs, often being mixed with ammonium nitrate for such applications.

Tetryl, $C_7H_5N_5O_8$, is formed when dimethylaniline is nitrated. It requires careful extraction and preparation for use as an explosive, as it is powdery and toxic. It has a detonation velocity of 24,000 ft/s (7300 m/s), a power value of 120, and a sensitivity of 70, which makes it a good intermediary charge to transfer a detonating shock wave from a detonator to a main charge.

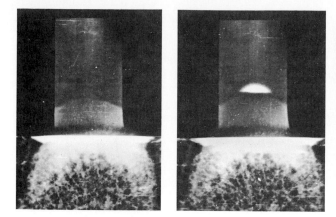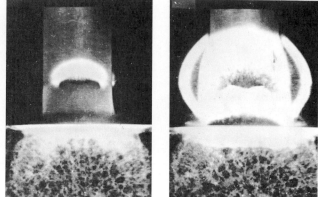

Above: A series of photographs showing the development of the detonation wave in a block of solid explosive, in this case a mixture of RDX and TNT. The series runs from left to right, and shows frames 2, 3, 4 and 7 of a high-speed sequence photographed with a time interval of two microseconds (millionths of a second) between each frame. Below: A composite metal plate, made of a sheet of carbon steel laminated with a covering of aluminum bronze. The sheets are placed slightly apart, explosive is spread across the bronze and detonated to force them together resulting in a permanent bond between the metals.

RDX ($CH_2N.NO_2$)$_3$, also called *Hexogen* or *Cyclonite*, was discovered by the German Henning in 1899. It is the product of the nitration of *hexamethylenetetramine*, and is a very powerful explosive (167) with a high velocity of detonation, 27,500 ft/s (8400 m/s), and a moderate sensitivity of 55. It has many uses in the military and civil fields.

Applications

Inevitably explosives find their principal applications in war, and modern propellants stem from developments made in the 1880s. In 1884 the French engineer Vieille gelled nitrocellulose with an ether-alcohol solvent, then cut the sheets into flakes, and named it "Poudre B" after General Boulanger. Nobel made *Ballistite* in 1888 by gelling nitrocellulose with nitroglycerine, and the English *Cordite* of 1889 was produced by Nobel's methods. As gun propellants these substances burn quite slowly, yet if the constituents are suitably ignited they detonate at many thousands of feet per second. The high explosives used for military purposes are mainly based on RDX and PETN with the most powerful ones being cast mixtures of RDX with TNT and aluminum. Plastic explosives are also based on RDX which is mixed with oils, waxes and pasticizers to give the required characteristics.

Shaping explosives can produce a range of useful applications. For example, a cone-shaped charge directs the explosive force accurately and concentrates it in a smaller area. This principle is used in artillery, demolition work and even the delicate task of piercing a hole in the casing of an unexploded bomb. Explosive forces can also be used for shaping metal or bonding metals together. The explosion forces the two surfaces of metal together under tremendous pressure, resulting in a permanent bond between them.

Explosives are widely used in mining but in cases such as coal mining, where there is danger from fire damp, the use is limited to permissible explosives. These are generally high explosives modified by the addition of common salt which reduces the ignition

Right: A plume of water created by the detonation of about 8.8 lb (4 kg) of explosives at a depth of 10 ft (3 m). Far right: Modern warfare techniques depend on a diverse range of explosives, each with specific components and uses. These can include anything from an RDX- and TNT-based high explosive designed to demolish a pinpointed area to a tiny antipersonnel mine used to maim rather than kill an enemy soldier.

risk. In addition, explosives such as Cardox are used; this consists of liquid CO_2 in a shell with a heater. On firing the gas is heated and expands rapidly to give an explosive effect. The Airdox and Airbreaker systems are similar but use very high pressure (10,000 lb/sq in.) compressed air.

Canal excavation, harbor deepening, and demolition are other fields in which large quantities of explosives are used. In seismic prospecting, for oil for example, shock waves generated by the detonation of a charge travel into the strata below and are reflected or refracted by the geological features of the area. By detecting these reflected waves a picture of the underlying geological formation of the site can be built up.

An unusual application is that of quenching oil and gas well fires. The largest ever, at Gassi Touil in the Sahara, which had burned for five months, was put out by Red Adair of Texas using 550 lb (249 kg) of dynamite, for a fee of $1,000,000.

Very high explosive powers are available from atomic reactions and some use is being made of this type of explosive in applications such as the production of underground storage caverns, encouraging the flow of oil from wells and deep seismic soundings. Proposals have also been made for the use of atomic explosives in major earthmoving works, such as the excavation of another Atlantic-Pacific canal in Central America. However, such applications are restricted by Test Ban Treaties which limit the amount of atmospheric contamination on Earth by banning any atmospheric and in-water atomic explosions.

See also: Ammunition; Bomb; Bullet; Grenade; Gun; Mine; Mining techniques; Quarrying.

• FACT FILE •

- Eighteenth- and nineteenth-century military siege techniques included the placing of heavy gunpowder charges in deep tunnels beneath battlements – known as mining. It was calculated that a mine 10 ft (3 m) deep in compacted earth would require 3000 lb (1360 kg) of powder, the equivalent of 90,000 musket cartridges. Charges up to 5000 lb (2268 kg) were common.

- In April 1947, 512 people were killed and 4000 injured by a tremendous explosion at the Texas City Refinery dock. The explosion took place on the freighter Grand-camp loaded with ammonium nitrate. This brought commercial interest in fertilizer-grade ammonium nitrate (FGAN) as an explosive when sensitized by a hydrocarbon, such as diesel oil. FGAN plus diesel oil eventually became the widest used, and cheapest, of all commercially available explosives.

- Modern antipersonnel mines contain tiny charges designed to maim rather than kill. The Spanish P4B mine used by the Argentines in the Falkland Islands contained no more than 6 oz (170 g) of plastic explosives, which could be detonated by a force of only 22 lb (10 kg) – about a tenth the weight of the average soldier.

Exposure meter

When taking a photograph, the photographer has to make sure that the correct amount of light passes through the lens aperture and shutter of the camera onto the light-sensitive film. If there is too little, the picture will be too dark, or *underexposed*; if there is too much, it will be too light or *overexposed*.

To accurately record a subject onto film, the photographer therefore uses an exposure meter, an instrument which measures the intensity of the light and either indicates or automatically produces the correct exposure for the picture to be taken.

In the early days of photography the correct exposure was found by trial and error, or by the photographer's own experience. Printed charts and tables were available which gave a rough guide to exposure in average lighting conditions.

A more accurate measuring system came with the invention of the *actinometer*. A light-sensitive paper was placed in the meter and put in front of the subject. By comparing the tone of discoloration caused by the light falling on the paper with a separate, preprinted tint card, timing the process, and referring to tables, an exposure was found.

This rather time-consuming method was superseded by the *extinction meter*, basically a tube with a dark interior and an aperture at each end. A transparent strip in blackness, and overprinted with black numerals, was placed between the ends of the meter. The subject was viewed through the tube and

a numeral found that, in relation to the illumination level, merged with its background tone. This number was transferred by calculation to give an approximate camera setting.

The photoelectric exposure meter was introduced in the U.S. in 1932, and is a far more advanced and accurate instrument. It measures the light level falling on a light-sensitive surface (a photoelectric cell) in electrical terms. In separate (off camera) meters, the photocell activates a needle over a scale of *exposure values*, which can be read out as a range of shutter and aperture settings. The photographer then adjusts the camera accordingly. When the meter is built into the camera itself, the needle can be designed to appear within the viewfinder. In this case in semiautomatic cameras, the cell is linked to either the aperture or shutter, leaving the operator one adjustment to complete the setting.

Types of cell

In the *selenium* cell, the light falls on a photocell which has two electrically conductive layers, sandwiching a layer of selenium. The light creates an electric potential between the two layers in proportion to its intensity, and this current is registered on a sensitive AMMETER by a needle. Because of their relative insensitivity, selenium cells are too large to be placed inside cameras and so cannot be used for

Below: High contrast poses particular difficulty for an exposure meter. Ideally it will be able to average the shadows U and the highlights O.

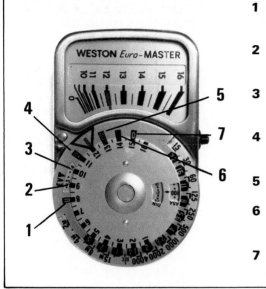

1 U position to determine exposure by measurement of darkest object

2 Quick-means facility to decrease exposure by three *f* stops

3 Quick-means facility to decrease exposure by two *f* stops

4 "A" position provides a means of selecting half exposure

5 C position gives double exposure

6 Quick-means facility to increase exposure by two *f* stops

7 O position to determine exposure by measurement of lightest object

Above: A traditional selenium cell exposure meter (left) and a newer, digital equivalent (right).

through-the-lens (TTL) metering. Their great advantage is that they require no external power source. The *cadmium sulfide* (CdS) cell is more sensitive to light than the selenium type, and can be used in duller lighting conditions. This cell is not electrically self-generating, however; the electric resistance of the cell varies with the intensity of the light, and a battery (usually a long-life mercury cell) is needed to power the circuit.

Cadmium sulfide exhibits the unfortunate property of remembering bright lights for several seconds after they are removed. Care should therefore be taken to avoid pointing a cadmium sulfide meter directly at a bright light source, such as the Sun, because this will lead to incorrect readings when subsequent measurements are taken. The silicon photodiode (SPD) is even more sensitive than the CdS cell and it responds very quickly to changes in light level. Many cameras include SPD meters which are fast enough to take an accurate reading in the few milliseconds between the lens stopping down and the shutter opening. Such cameras may adjust their shutter speeds to take account of minor errors in the stopped-down aperture of the lens. SPDs are also fast enough to be used for metering light from flash guns. The principal disadvantage of most SPDs is that they are relatively insensitive to blue light. This may be corrected by fitting a blue-tinted cover to the SPD.

Using a meter

An exposure meter is not infallible, and unless used correctly, will give inaccurate results. The most common method of using the meter is to gauge a general or overall level of brightness. This will take into account the darker and lighter parts of the sub-

ject, and give an average reference for exposure. Having been set for the *speed* (sensitivity) of the film in use, the meter is pointed toward the subject and an average reflected light value is read off.

Another method is to use an *invercone*, a translucent attachment which is placed over the photocell. This reduces the reading somewhat but instead of measuring the reflected light, the *incident* light falling on the subject is used. This method overcomes problems caused by one part of the picture being much brighter or darker than the subject.

The exposure value (EV) is a combination of shutter speed and aperture, and its choice depends on the depth of field and movement of a particular subject. For example, the camera may be marked in apertures (*f* numbers) between *f*3.5 and *f*22, and have shutter speeds of between one second and one thousandth of a second. The meter will indicate a number of these combinations, all theoretically correct (*f*8 at 1/500 second being equivalent to *f*11 at 1/250, since although the speed has been halved, the aperture has been made smaller accordingly).

Meters read a particular area of the subject in front of them. This is the *acceptance angle*, and will differ between models. More advanced meters may read a general area, and also have the facility to take a spot reading of a small part of the subject. Special meters and attachments are available for obtaining light readings in inconvenient places, such as at the ground-glass screen of a view camera and on part of a microscopic specimen.

See also: Ammeter; Camera; Film; Light; Shutter.

Extrusion

Extrusion is a useful process for shaping both metal and plastic products, notably rods, tubes and various solid and hollow shapes, by forcing them through a die, in much the same way as frosting is forced from the nozzle of cake decorating equipment.

Plastic extrusion

Plastics are extruded by being forced through orifices or dies which give them a particular shape. Flat sheet, pipe, tubular film (plastic bags), coated wires, and shapes such as curtain rods and rings are some of the many products manufactured by this means. The plastics used are predominantly of the thermoplastic type, which includes polyethylene, polypropylene, polystyrene, polyamide, polyvinyl chloride and many more. Such materials are extruded in the highly viscous molten state, typical processing temperatures being in the range 302 to 572° F (150 to 300° C).

Extrusion machines must be capable of both melting the plastics and generating extrusion pressures as high as several hundred atmospheres at the dies. They also have to perform other functions such as the mixing in of coloring matter and other additives, occasionally reinforcing agents such as glass fibers. Another important requirement is for continuous steady operation.

The most commonly used machines are single

Right: Helical extrusion is used for reducing billets directly to wire. In one operation it combines hydrostatic extrusion, normal extrusion and an intermediate stage similar to lathe turning.
Below: A typical plastics extruder designed to handle rigid cellular PVC. Powdered PVC is fed into the machine. It is then melted by heating, forced through a die and cooled to form baseboards.

screw extruders. The plastic material, normally in granular or powder form, is fed via a hopper to a coarse-pitched Archimedean screw rotating inside an electrically heated barrel.

The screw generally has three distinct sections. In the feed section, which conveys solid material from under the hopper, the screw channel depth is constant and relatively large. The subsequent compression section has a uniformly reducing channel depth and is intended to compact the plastic and force it into contact with the barrel to encourage melting. Melting is achieved mainly by a combination of heat conducted from the barrel and heat generated by intense shearing in the molten layer formed between the barrel and the solid material. As its name suggests, the third or metering section of the screw, in which the channel depth is constant and relatively small, is intended to control the output from the extruder in terms of quantity, steadiness and homogeneity. The relative motion between the screw and barrel creates an efficient mixing action in the now molten plastic, and is capable of generating the pressures required for extrusion.

Extruders range in screw diameter from about 1 to 12 in. (25 to 300 mm), and give outputs from a few

pounds to several tons per hour. Barrels and screws are made from hardened steel to minimize wear, since for efficient operation the clearances between them must be small. Screws are normally run at about 100 rpm, driven by electric motors via reduction gears. Any air entering with the solid feedstock is forced out by the subsequent compaction. Some plastics tend to contain volatile contaminants such as unconverted monomer from the polymerization process, so modified screw designs are used which allow the use of a vent about half way along the barrel. There is normally a filter for solid debris between the screw and die.

The screw extruder also forms the basis of several other plastics-processing machines, notably those for blow molding and injection molding. It is therefore the most important single piece of equipment in the plastics industry.

Metal extrusion

In this process a billet of material is forced under compression through an orifice which has a much smaller cross-section than that of the billet. The original process has been developed into a variety of loosely related methods.

An essential feature of the extrusion process is that the *shear* forces needed for the plastic deformation of the material are developed by the compression of the material. It is the lack of tensile stresses that allows even brittle materials to be extruded without breakages occurring. The best-known process is *direct (solid)* extrusion. The billet, typically 6 in. (150 mm) diameter by 18 in. (500 mm) long, of squat cylindrical shape, is inserted into a strong steel container in which it fits quite closely. At the front end of the container is the die which contains one or more orifices to determine the cross-sectional shape of the extruded material and which, in the case of conventional forward extrusion, is fixed in relation to the container. The back end of the container is closed by a ram which moves forward to apply the extrusion force to the billet. The metal is therefore forced through the die, whereas in *reverse extrusion* the die is not fixed to the container but is forced into the metal billet by the ram.

Extrusion was first used for the manufacture of lead pipe. When the process was later extended to metals of higher melting point it was developed as a hot working process, probably because it was thought that these harder materials had insufficient plasticity at room temperature. Thus, copperbase materials are commonly extruded at temperatures of about 1290 to 1650° F (700 to 900° C) while the corresponding figures for aluminum alloys are 750 to 1020° F (400 to 550° C). The hot extrusion of steel requires temperatures of 2010 to 2190° F (1100 to 1200° C) but these high temperatures are only made possible by the use of glass as a

Above: Apart from wire and rods, a wide selection of metal blanks can be readily formed by the various extrusion processes.

lubricant to protect the tools. It was later realized that cold extrusion of many hard materials is quite practicable provided a suitable press of sufficiently large capacity is available. Recent developments, however, include the use of induction heaters and hydrogen-atmosphere furnaces to allow the extrusion of molybdenum at up to 3270° F (1800° C).

The cold extrusion of difficult materials is facilitated by hydrostatic extrusion. In this process the billet is completely surrounded by a fluid which transmits the force from the ram hydrostatically, and also acts as a lubricant.

The required extrusion force is determined by the cross-sectional area of the billet, the resistance to deformation of the material and the extrusion ratio. Large forces are required and presses have been constructed with capacities in excess of 20,000 tons, though capacities around 2000 tons are normal.

In impact extrusion, a metal slug is positioned in a die and struck with a shaped punch that has a clearance in the die, the force of the blow extruding the metal into the clearance space between the die and the punch. This process gives good mechanical properties to the metal and it is possible to produce complex parts in a single operation. A common example of an item produced by impact extrusion is a metal toothpaste tube.

Usually metal extrusion is a batch process with billets being extruded one at a time. The Conform process uses a rotating wheel to force solid or granular feedstock into a die chamber and through the die to give a continuous process.

Foodstuffs

The extrusion process is also widely used in the manufacture of foodstuffs, such as spaghetti and corn snacks, and is finding increasing application in the processing of artificial foods.

See also: Automobile; Plastics production.

Eye

The eyes receive visual information about an organism's surroundings and pass it to the *visual cortex* in the brain for analysis and interpretation. Only a small area of the eyes is visible – the bulk is deeply set into the skull in hollows called *orbits*. For added protection, the eyelids can be closed voluntarily, to protect against strong light, for example, or involuntarily, when there is sudden movement near the eye, say.

Blinking cleans and lubricates the eyeball by distributing fluid from the tear glands and *sebaceous* (fatty secreting) glands at the edges of the eyelids. The eyelids are lined with a thin layer of epithelial (skin-type) tissue, which is continuous with the front of the eye and prevents objects, such as dust and contact lenses, from slipping behind the eye. The eye can blink automatically and quickly when danger, such as a bug, approaches.

The tear glands are not strictly part of the eye. They are in the top part of the orbits and secrete a solution of sodium hydrogen carbonate and sodium chloride which moistens, lubricates, and cleanses the exposed area of the eye. The tear solution also contains lysozyme, an enzyme which helps to prevent infection.

The eyeball

The eyeball is completely enclosed by the *sclera* or white of the eye – a tough, opaque outer layer made up of inelastic tissue which keeps the eye in shape. The front of the sclera is the *cornea*, a transparent disc which lets in and refracts (bends) light. Stimulations of nerve endings in the cornea cause tear production and reflex blinking, and also make the cornea sensitive to pain.

The inside of the sclera, but not the cornea, is lined by the *choroid*, a pigmented layer with many blood vessels. The black pigment reduces reflection of light inside the eye. At the front, the choroid is modified to form the iris, which is opaque and pigmented. The pigmentation ranges from none, as in pink-eyed albinos, to dark brown.

The iris opens and closes to vary the size of the pupil, which lets light through into the eye. The movement is controlled by involuntary muscles in the iris which respond to light levels; a high level of light causes the pupil to contract and restrict the amount of light passing through, and a low-intensity light causes the pupil to dilate to maximize the amount of light gathered.

The eyeball is divided into two parts – the anterior (front) and the posterior (rear), which is three times larger. The anterior eyeball contains the *aqueous humor*, a thin, liquidy blood plasma which

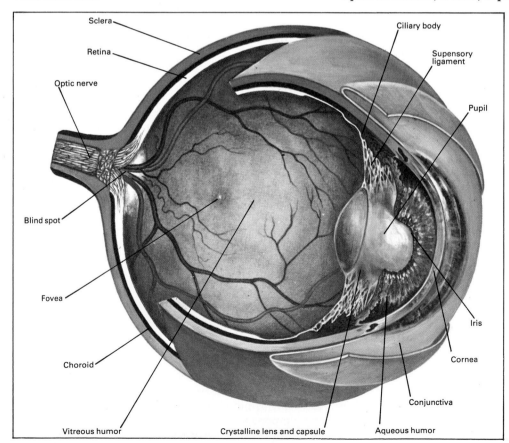

Left: The eyeball, together with the muscles that control eye movement, blood vessels, nerves and the lacrimal gland are lodged in a bony socket in the skull. The pupil determines the amount of light let in to the eye by the contraction of the iris. Behind the iris, the aqueous humor nourishes the cornea and has little effect on light rays; it is renewed every four hours. The light rays pass through the lens, which refracts them to focus on the retina.

Sclera

Retina

Optic nerve

Blind spot

Fovea

Choroid

Vitreous humor

Crystalline lens and capsule

Aqueous humor

Conjunctiva

Cornea

Iris

Pupil

Supensory ligament

Ciliary body

Right: In bright light the muscles of the iris contract, causing the pupil to shrink. Far right: In dim light the iris relaxes and so allows the pupil to expand.

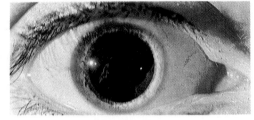

Above: Owls have tubular eyes which are deeper than they are wide. The cornea is huge and the iris is capable of immense expansion and contraction, enabling owls to hunt at night.

contains the only supply of oxygen and nutrients to the lens and cornea. The aqueous humor is produced by the *ciliary body*, a thickened part of the choroid containing muscles, large blood vessels, and glands. The aqueous humor exerts a pressure of about 0.5 psi (25 mm Hg) to keep the cornea and lens in shape.

The lens itself divides the anterior and posterior parts of the eye. It is a biconvex, transparent body attached by the suspensory ligament to the ciliary body. The lens further refracts the light and focuses it on the retina. The muscles in the ciliary body alter the shape of the lens, enabling it to focus on near or distant objects. The lens is made up of thin fibers and epithelial cells in concentric layers like an onion. The refractive power of the lens increases toward the center.

The space behind the lens is filled with the jelly-like *vitreous humor*, which maintains the shape of the eye and helps to further refract light. Running through the vitreous humor from the back of the eye to the lens is the *hyaloid canal* which, in the embryo, contained a blood vessel supplying the lens. This vessel disappears after birth.

The eye and vision

If the eye is thought of as a camera, then the retina would serve as the film. The retina is a layer of light-sensitive cells on the inside of the eye. In effect, it is a continuation of the optic nerve, which enters the eyeball at the back through the sclera and choroid. There are two types of cells – *rods*, which are sensitive to black, white and shades of gray, and *cones*, which react to color. There are about twelve million rods and seven million cones.

Rods contain a pigment (rhodopsin or visual purple) made up of retinene (obtained from vitamin A) and opsin. Light causes the rhodopsin to decompose into its constituent parts and stimulates the nerve leading from the rod into the optic nerve and then to the visual cortex, where rhodopsin re-forms. Very little light is needed to stimulate the rods, so they are particularly important for night vision. The cones, however, need high levels of light to work, so at night colors appear muted or absent.

There are three types of cones – the first is sensitive to red, the second to green, and the third to blue. Colors are perceived according to which cones are stimulated by the wavelength of the light entering the eye. Cones are concentrated in one particular part of the eye – the *macula* – a small hollow in the middle of the retina and the point at which most light falls when the eye is at rest.

The center of the macula, the *fovea*, is packed with cone cells and is the area of sharpest vision. Cones diminish in concentration toward the edges of the retina. There are no light-sensitive cells at the point where the optic nerve leaves the eye; this is termed the blind spot. It does not appear as a blank in the image, because one eye compensates for the other, and the image is projected onto the fovea.

Light is focused on the retina by *accommodation*, which varies the shape of the lens. To view a distant object, muscles in the ciliary body relax, tightening the suspensory ligament and stretching the lens to make it thinner. When viewing a close object, the

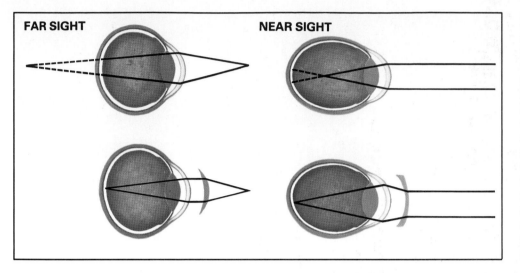

FAR SIGHT   **NEAR SIGHT**

Far left: Far-sightedness is the inability to focus close objects on the retina, due to the eyeball being too short. Left: Near-sightedness occurs when the eyeball is too long and light rays from a distant object are focused in front of the retina. Glasses correct both conditions by focusing the image on the retina.

ciliary muscles contract, the suspensory ligament relaxes and the lens thickens and shrinks.

Light from an object falls on the retina as a focused image, which stimulates the rod and cone nerve cells. This image is smaller than life size and upside down. The stimulations travel to the brain, which perceives the image the right way up and the correct size. Together, the two eyes send the brain a stereoscopic view which gives the image depth and helps the brain to judge distance.

Quantifying vision

Vision is measured against a theoretical normal eye rated as 20/20 or 6/6. An eye with 20/20 vision can see at 20 ft (6 m) what the normal eye can see at 20 ft (6/6 is the same formula expressed in meters). Two 20/20 eyes, and the ability to see a three-dimensional image, is normal. In practice, eyes can be above average. For example, a 20/13 (6/4) eye can see at 20 ft (6m) what a normal eye can see if it is within 13 ft (4 m). Usually, adults' eyes are below average. A 20/30 (6/10) eye, for example, has to be at 20 ft (6m) before it can see what a normal eye can see from 30 ft (10 m).

Common eye defects

The most obvious eye defect to the onlooker is *strabismus* or a squint. Each eyeball has six muscles between the sclera and the back of the orbit. These muscles are arranged in opposing pairs – the superior and inferior rectus, which make the eye look up and down, the superior and interior oblique, which rotate the eye, and the medial and lateral rectus, which turn the eye from side to side.

The brain coordinates these muscles so that they move together. In some individuals they are not properly coordinated and one eye may "point" in an entirely different direction from the other. If uncorrected, this condition is serious, especially in the very young.

Myopia (near-sightedness) and *hyperopia* (far-sightedness) are caused by a slight defect in the shape of the eyeball. In myopia, the light is focused at a point in front of or short of the retina, blurring the image on the retina. In hyperopia, the image reaches the retina before it is focused. Effectively the image is focused behind the retina – the focus is too long and the image is blurred.

Astigmatism is a defect in the cornea. If the cornea is not perfectly hemispherical, light rays do not come together at a single point on the retina, as they should, and there are two points of focus, causing blurred vision.

Below: Several features protect the eye. Eyebrows stop sweat from running into the eye, eyelashes keep the eye clean, eyelids sweep dirt from its surface and tears bathe it.

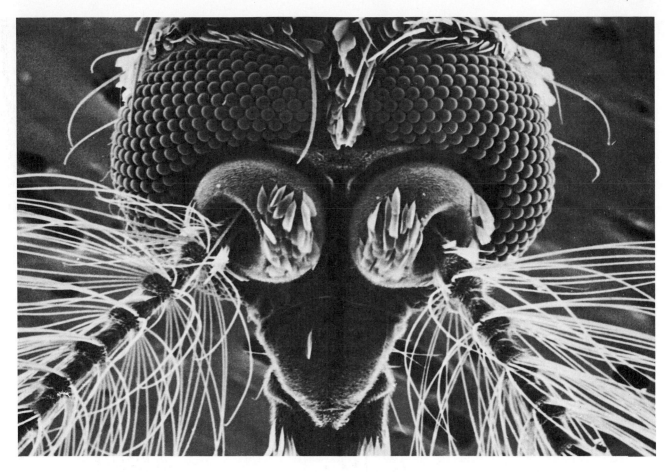

Above: Insects have compound eyes with hundreds of individual *ommatidia* or optical units, as with this mosquito. It is uncertain how much insects can see but some distinguish colors.

Another eye defect is *presbyopia*, the far-sightedness associated with aging. It is caused by a loss of tone in the muscles controlling the lens. As the muscles age, they become less efficient at accommodating the eye, so objects have to be held further away from the eye to bring them into focus. All these defects are correctable, usually just by wearing glasses or contact lenses.

Nonhuman eyes

The eyes of vertebrates are similar, differing only in small details, such as the shape of the lens. Within the vertebrates, the main difference is between primates, who can see color, and the rest, who are thought not to be able to see color.

Some creatures, for example, birds, cats, rabbits and alligators, have a third eyelid, the *nictitating membrane*. This is transparent and can close over the whole of the exposed part of the eye, cleaning and protecting with little or no loss of vision. When closed, the membrane is situated in the bottom corner of the eye nearest the nose. In humans, it is vestigial.

Compound eyes

Many kinds of arthropod (including shrimps, spiders and insects) have compound eyes. These are hundreds of individual *ommatidia* or optical units, which refract light and pass information to the optic nerve. The eye is roughly hemispherical. On the surface is the cornea equivalent – the *cuticular lens*. Below this are the *vitrellae* – four cells with transparent, refractive inner edges which form a crystalline cone. The cells below this are the *retinulae*. These, too, have refractive inner edges and contain the visual purple. They are called the *rhabdom*.

The retinulae are the receptor cells and project down into an optic *ganglion*, which is connected to the large optic nerve that goes to the brain. A large number of facets in a compound eye gives it a large field of vision. The extent to which creatures with compound eyes see is uncertain but probably some can differentiate shapes and color. The main function of this type of eye seems to be to distinguish between intensities of light.

Many arthropods have *ocelli* or simple eyes. These are light-sensitive spots.

See also: Brain; Enzyme; Glasses; Light.

The surgeon's eye

Many visual problems, especially in elderly people, are caused by disturbances of the blood vessels at the back of the eye. But gaining access to these blood vessels to investigate the nature of the disorder has proved extremely difficult. The technique of *fluorescein angiography* now enables the specialist to examine the eyes with relative ease. It takes advantage of the physical property of fluorescein in absorbing light of one wavelength and emitting it at another wavelength. Blue light is shone onto blood vessels carrying fluorescein and, by monitoring the resulting emission of green light, the path of fluorescein through the blood vessels can be followed.

Having identified the site of the problem affecting the retina, the eye surgeon is faced with the problem of treating it. Obviously, direct access to these delicate tissues is difficult, but they may be treated with minimal interference, with a laser.

The application of laser technology to the treatment of retinal disease is simple. The patient's pupil is first dilated and then the external surface of the eye is anesthetized with drops. Next, the surgeon fits the eye with a small contact lens, equipped with three mirrors at varying angles of inclination. These enable the surgeon to focus on any particular part of the retina.

With the patient seated comfortably at a microscope attached to the laser machine, the surgeon can fire the laser at the diseased area of the retina causing a minute, highly localized retinal burn. As many as 500 such tiny burns may be used in a single treatment, and treatments may be repeated.

Naturally, this laser burn causes destruction of both normal and abnormal retinal tissue. This means that pathological lesions affecting the optic nerve or the *macula* cannot be treated directly. This excludes using lasers to treat many eye diseases in the elderly patient, but the technique is invaluable for diabetic patients, often enabling them to lead an independent life instead of being condemned to the limitations of blindness.

Cataract is the condition in which the lens of the eye becomes progressively clouded. Most usually seen in the elderly, its removal is the most common operation that the ophthalmic surgeon performs.

By breaking up the cataract within the eye (*phako-emulsification*), and removing it by suction (*aspiration*) through a small incision, the surgeon can safely perform cataract surgery with only a 24-hour stay in a hospital for the patient.

In the *phako-emulsifier*, an acoustic vibrator contained in the tip of a small ultrasonic probe oscillates back and forth about 40,000 times per second,

Below: An eye surgeon uses a laser to treat a patient with a diseased retina. The laser beam burns tiny holes in the diseased area.

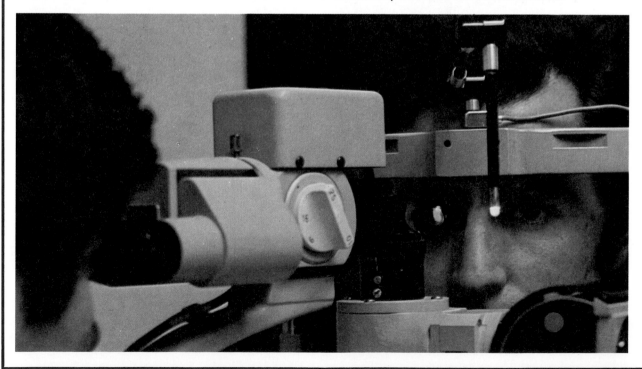

REMOVING A CATARACT

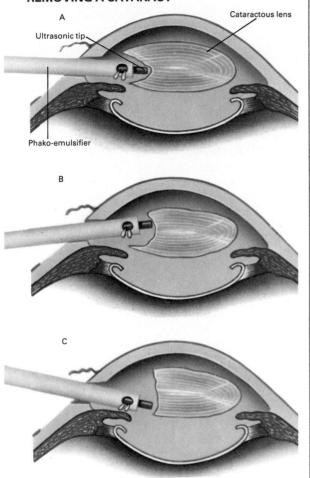

Above: Phako-emulsification has revolutionized cataract surgery, reducing hospital stays from five days to twenty-four hours. The clouded lens is fragmented in stages by the ultrasonic vibrations of the emulsifier. First, the equator of the lens is attacked (A); then the upper shelf is removed (B); next the lower shelf is taken out (C) and so on.

causing the lens to fragment. Since this process generates heat, it is combined with a cooling system. The ultrasonic probe fragments the cataract but does not remove it from the eye. So it must be used in conjunction with an *aspiration system*, which sucks out the fragmented cataract, and an *irrigating system*, which prevents disintegration of parts of the eye. These three systems can be combined in a small hand-held probe which is inserted into the eye through an incision.

The initial incision is made with a diamond knife

at the junction of the transparent cornea and the opaque sclera. Then a sharp hook, attached to the irrigating system, is inserted into the anterior chamber of the eye and used to tear a large gap in the anterior capsule of the cataract. The hook is then removed and replaced by the ultrasonic handpiece which fragments the cataract and aspirates the debris at the same time.

This technique removes the cataract but leaves its posterior capsule intact. It is known as *extracapsular extraction*, as distinct from *intracapsular extraction*, in which the posterior capsule is removed. The advantage of the former technique is that the remaining posterior capsule provides mechanical support for the vitreous gel of the eye, lessening the long-term risk of retinal detachment and cystoid macular edema – two complications leading to serious visual disturbance.

Older techniques of extracapsular extraction failed to remove all the cataractous material, which resulted in inflammation of the eye and an increased incidence of complications. By using the irrigating-aspirating system of the probe under microscope control, all the cataractous material can be removed, leaving the eye uninflamed.

Removal of the cataract does not, on its own, restore vision, since the lens was in the patient's younger days the main focusing mechanism of the eye. To bring back sight after the operation an optical device must be used. There are three main types: cataract glasses, contact lenses and intraocular lenses.

The disadvantage of glasses is that they cause 25 per cent magnification of objects and a constricted field of vision. Although contact lenses give a much better quality of vision, learning to wear them for the first time at the age of 80 is a daunting task. A solution to the problem is to insert a replacement lens into the eye at the time of removal of the cataract – this gives good vision.

All these lenses consist of a central optical device – usually a planoconvex lens – with varying types of support on the sides to fix the lens in position in the eye. The lens is made of polymethylmethacylate (PMMA), which is inert within the eye.

Naturally, the presence of an artificial lens within the eye carries the risk of infection or other complications (though this is only slight), whereas wearing cataract glasses involves no such problems. Set against this slight risk, however, is the fact that an elderly patient coping with the difficulties of wearing cataract glasses is probably likely to have problems balancing and judging distances, and is thus more likely to end up in an orthopedic ward with a fractured femur.

Fabric printing

Fabric printing is almost wholly mechanized, having been developed from methods such as block printing (which is similar in principle to printing from a lino-cut) or printing by means of a stencil. Block printing is an expensive process, often carried out by hand, but can offer very high quality.

Probably the most common printing method is by means of rollers, the metal surfaces of which are engraved with the design. Each roller is *furnished* with color by means of a brush-covered furnishing roller which in turn dips into a trough of color paste. Surplus color is scraped off the roller by an accurately ground *doctor blade*, leaving the color in the recessed engraved design portions. The color roller presses against the fabric, which passes over a large rotating cylinder. A second doctor blade scrapes off any fibrous impurity from the roller face before it is refurnished with color. A separate roller is required for each color component of the design; thus an

Right: The word batik is used for a wax-resist method of dyeing. Designs are produced by covering each part of the cloth in turn with wax.

Below: Hand silk-screen printing at a textile mill in Bombay, India. The dye is forced through the holes in the screen by a squegee.

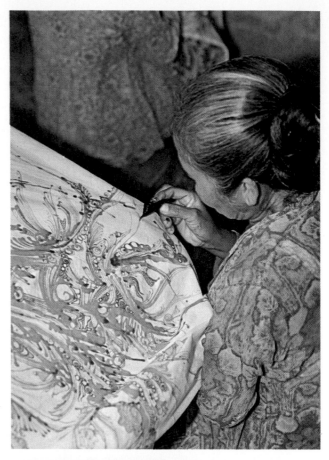

Above: Dyes used in fabric printing are thickened to the consistency of paste to provide sharp outlines. The roller engraved with the design is below, above it the furnishing roller. Excess paint is scraped off by the doctor blade.

Right: Fabric printing. Dye is taken up through the furnishing and continuous printing roller. The blanket gives a greater resilience.

eight-color print requires eight engraved rollers, which are expensive. The whole machine is of massive construction, and setting up the heavy rollers for a printing operation takes time. Accordingly, the method is most suitable for long printing runs of many thousands of feet. This process can be used to print either one or both sides of the fabric.

Screen printing

In screen printing, a stencil is developed photochemically on a strong fine-mesh material which has been stretched over a large rectangular frame. This material was originally of silk – hence the name silk-screen printing – but is now almost wholly made of tougher materials such as nylon, polyester or metal. The screen lies on top of the cloth to be printed, the color being forced through the open parts by a squeegee. In the manual system, now used only for specialty work, the cloth is stuck to a long table and the operators move the screen along – one screen for each color component. In the mechanical systems, the cloth is carried on an endless belt over a short table; the screens are lowered,

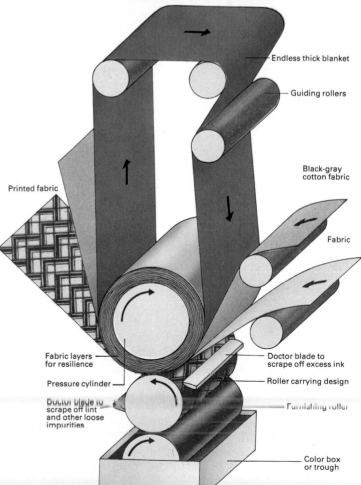

Endless thick blanket

Guiding rollers

Black-gray cotton fabric

Printed fabric

Fabric

Fabric layers for resilience

Pressure cylinder

Doctor blade to scrape off excess ink

Roller carrying design

Doctor blade to scrape off lint and other loose impurities

Furnishing roller

Color box or trough

Above: Patterns for machine-programed film strip, used in automated fabric printing.

the printing stroke made, and the screens lifted to permit the cloth to move on the distance of one pattern repeat. Production is slower than by roller, but as the screens are cheaper to make and take less time to set up, the method is particularly suitable for short runs. It will also permit wide design repeats, but is less suitable for unbroken patterns; for example, it would be impossible to print a lengthwise stripe without visible overlapping joins.

Transfer printing

The newest method, which though forming only a small fraction of total printwork is expanding rapidly, is that of *transfer* or subliastic printing. Paper, which because of its rigidity and smooth surface can be roller printed very accurately, is printed with a type of dye which vaporizes (sublimes) when strongly heated to approximately 390° F (200° C). If this paper is then passed through an accurately controlled hot press along with the textile fabric, dye transfer takes place. No aftertreatment is necessary as only the dye is transferred, and there are fewer problems with fine nonabsorbent fabrics which are difficult to print conventionally. Some limitations arise in respect of dye fastness, and only fabrics made from fibers capable of being dyed with this particular type of dye can be printed in this way.

Dyes and pigments

Colors used in printing commonly comprise dye solutions thickened to the consistency of a thin porridge by means of starches or gums. The product is a color paste and permits sharp outlines because the color does not spread. After printing, the fabric is dried and the color *fixed*, usually by passing through a large steam chamber. Some combinations of dye and fiber require high-pressure steam to fix them. Some dyes must be oxidized to obtain the correct shade, and the fabric must be thoroughly washed to remove the thickener together with surplus dye. These aftertreatments are of vital importance for proper color values and fastness.

Pigment printing differs in that it utilizes a pigment which by itself would have no affinity for the fabric but which is applied along with a resin, firmly binding it to the fiber. Pigment and resin are thickened not with starches but with a spirit emulsion the consistency of thick cream. There is no inconvenient steaming or afterwash, a simple heat treatment, for example, four minutes at 266° F (130° C), being all that is required. The method is not, however, suitable for all types of fabric.

As well as applying color in the form of a pattern (the *direct* style) it is possible to produce patterns removing the color from fabric which has previously been dyed in the conventional way. By printing with a paste containing an appropriate chemical, the dye is destroyed.

See also: Clothing manufacture; Dyeing process; Fiber, synthetic; Ink; Printing.

Farming, intensive

Intensification in livestock farming improves productivity of meat, eggs, and milk by increasing the numbers of animals supported by available land, reducing the amount of labor required for management, and increasing the productivity of the individual animal. But when the numbers are increased it creates problems: livestock can no longer support themselves by grazing or foraging, and must have expensive supplements or manufactured feeds. Disease can be a potential problem, and diets and housing techniques are often adapted to minimize this. Housing has the added advantage of making the operation independent of the seasons and weather, allowing continuous production.

In all intensive systems, livestock producing eggs or milk, or being fattened for meat production, are completely separated from the breeding stock, which are usually kept under less intensive conditions. Maximum meat productivity is achieved by selective breeding to produce as many offspring as possible with rapid growth rates and good feed efficiency: that is, to put on the maximum possible bodyweight for each unit of feed consumed.

Growth rates slow as animals near maturity, and livestock are usually slaughtered while still relatively young, as to raise them until they are fully

Below: Pigs being intensively raised in pens in a controlled environment. Lactating sows have to be prevented from rolling onto their piglets.

grown would require a disproportionate quantity of feed. For example, a broiler chicken reaches a weight of 4 lb (2 kg) in 7 weeks, when it is normally killed. For each 2¼ lb (1 kg) of feed it consumes, it gains 1 lb (0.45 kg) in weight. If it is allowed to grow to a large capon, however, it will soon take 3 lb (1.4 kg) or more of feed to put on an additional pound or half kilogram of weight.

In animals growing at such a rapid rate, nutritional requirements vary as the animal grows, and very careful control of feed ingredients is necessary. Growth rates may be further stimulated by supplementing the diet with growth-promoting substances, such as antibiotics for use in animals. Similarly, efficiency of production of eggs or milk is measured in terms of the quantity of feed consumed.

In most intensive farming systems, livestock are kept in sophisticated, controlled-environment houses. In traditional farming, solid and liquid waste are naturally dispersed over large areas of pasture, but with intensification, vastly increased quantities of waste are generated within a restricted area. When possible, automated disposal systems remove wastes, which are often stored in lagoons to be broken down by bacteria, dried and burned, or used as fertilizer.

Broiler chickens
This is the most highly specialized and efficient form of intensive farming; four billion birds are produced annually in the U.S. Day-old chicks are raised in large houses with controlled ventilation, heating and lighting. Feed is usually supplied auto-

matically by a conveyer system, and water is piped in. A typical modern broiler unit consists of several adjacent poultry houses, each containing 15,000 birds in large pens. The entire flock is supplied from the hatchery as day-old chicks, already vaccinated against some common diseases, and the birds may not be handled again until they are collected for slaughter. Because of extensive automation, the entire unit can be operated throughout the growing period by a manager and two assistants. To prevent disease, broilers are usually fed a medicated diet throughout their lives.

Battery chickens

The majority of eggs produced come from birds in battery houses. From three to six pullets are placed in each battery cage just as they begin to lay. Feed is supplied by conveyer or moving hopper, and water by a nipple drinking system. Droppings pass through the wire cage floor, usually either into a droppings pit or onto a moving belt which removes them from the house. As eggs are laid, they roll through a gap at the cage front, and are collected by hand or by conveyer. Light affects the hens' production of hormones involved in egglaying, so adequate house lighting is desirable. Due to the wire floors, contamination of the eggs is minimized. Medicated feed is not desirable for hens producing eggs

Above: An intensive veal production unit with surplus male calves from dairy breeds. Three million veal calves are produced in the U.S. annually.

because drugs may collect in the eggs. The threat of disease is reduced by keeping the birds away from possible infection from their droppings.

Beef and veal production

In North and South America, intensive beef production is based on vast outdoor feedlots which may contain several thousand beasts. The cattle are beef breeds, such as the Hereford, or hybrids obtained by crossing milk cows, Holsteins, with beef breeds, Hereford or Angus. Feed based on corn is supplied from trucks fitted with hoppers, which discharge into feeding troughs as the vehicle drives along the feedlot. Alternatively, the corn may be fed as silage, discharged by augers from large silos direct to the feed troughs. Beef cattle grow to 900 to 1100 lb (400 to 500 kg) in less than 12 months.

Intensive veal production is now predominantly based on large indoor pens. Surplus male calves born to dairy breeds such as Holsteins are fed on a liquid milk-replacer diet from multiple station feeders. Some veal calves are housed in individual pens on the smaller-scale production units. Three million veal calves per year are produced in the U.S. Other intensive systems are used with dairy herds where milking and feeding are partly automated in the zero-grazing system.

Hog production

Hogs are usually raised intensively in small groups in pens. Production techniques vary according to the feed materials used. Lactating sows are frequently restrained to prevent piglets being crushed. Hogs grow to an average liveweight of 240 to 245 lb (about 110 kg).

See also: Agricultural technology; Fertilizer; Fishing industry; Hydroponics.

• FACT FILE •

- To force winter salads, the Romans used an early form of greenhouse farming, using panels of translucent mica. The Emperor Tiberius ate out-of-season cucumbers daily at his Capri retirement palace. They were force-grown under mica cloches.

- In World War II, the U.S. Air Force installed soilless culture to provide fresh vegetables at remote air bases. The first air force hydroponic farm was on Ascension Island, where special beds irrigated with distilled seawater provided 94,000 lb (42,600 kg) of tomatoes, cucumbers, lettuces, radishes and green peppers in the first year.

- China leads the world in fish farming, producing almost 50 per cent of the annual international harvest. Fish farming produces some 7 per cent of all fish supplies globally, and the intensive methods used include the feeding of high protein diets, and increased tank water temperatures.

Fat

Fats are an important part of a living organism's diet, providing twice as much usable energy, weight for weight, as glucose. They are found in many foods, notably meat, milk products (such as butter and cheese), nuts, and some fruits, such as olives, avocados and bananas.

Chemical composition

Chemically, fats are compounds of one molecule of glycerol (glycerine) and three molecules of a fatty acid. The chemical formula for glycerol is:

$$\begin{array}{ccccccc} & OH & & OH & & OH & \\ & | & & | & & | & \\ H - & C & - & C & - & C & - H \\ & | & & | & & | & \\ & H & & H & & H & \end{array}$$

Fats are formed when the three hydroxyl (OH) groups are replaced by fatty acids. The fat produced depends on the fatty acid that combines with the glycerol and whether all OHs are replaced by the same fatty acid. For example, in the fat tripalmitin – $C_3H_5(C_{15}H_{31}COO)_3$ – the same palmitic acid radical ($C_{15}H_{31}COO$) replaced the hydroxyl groups. This fat is termed a simple glyceride. Most natural

Below: Though fat acts as an effective insulator, obesity is a common problem in the West today.

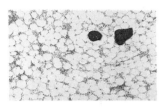

Left: Adipose tissue is the body's energy store. Fat accumulates in droplets and completely fills the cells. Women tend to have more fat on them than men.

fats, however, are mixed glycerides, combining radicals from two or three acids.

Fats are often referred to as being saturated or unsaturated. Saturated fats have a greater proportion of hydrogen atoms than do unsaturated fats. Generally, fats that contain greater amounts of saturated fatty acids are solid at normal temperatures, whereas those with greater amounts of unsaturated fatty acids tend to be much softer or even liquid, as is olive oil. In fact, the distinction between fats and oils is largely physical: glycerides are termed fats if they are solid at 68° F (20° C), and oils if they are liquid at this temperature.

Fat and energy

In the body, fats are digested with the aid of enzymes called *lipases*, which break them down into the component fatty acids and glycerol. Some of the fatty acids are used at once to provide the body with energy and others are reconverted into fats and stored in specialized fat cells. In warm-blooded animals, these fat cells accumulate around the major organs, particularly the heart and kidneys, and in the mesentery – the lining membrane that

Left: Fat people often have fat children. Scientists suspect a fatness gene may be inherited, though passing on family eating and exercise habits may also contribute. Right: A giant mincer making soap noodles. Soaps are made by a reaction of fatty acids with caustic soda, and recirculated lye. The particular fatty acids used determine the soap properties.

holds the intestines in place. Fat cells are also deposited in large quantities in the *dermis* (the deep layer of the skin) as *adipose* tissue, where it provides an effective insulator against heat loss.

In humans, skin fat can be responsible for up to one fifth of body weight, particularly in females, who have a thicker, better-developed fat layer. Marine mammals, such as whales, have an enormously thick layer of adipose tissue (blubber) to protect them from the cold sea. Warm-blooded animals that hibernate lay down large reserves of fat while they are active and this keeps their bodily functions working throughout their winter sleep.

In animals with a variable blood temperature (cold-blooded animals), body temperature is dependent on the ambient air or water temperature. The need for such animals to conserve body heat is not so great, so they store fat in the liver rather than under the skin, and in the fat bodies or fingerlike structures in the abdomen.

Fats, particularly animal fats, are important sources of the fat-soluble vitamins A, D, E and K. Vegetarians whose diet includes no animal products may need to take vitamin supplements.

Obesity

The Western diet is typically high in animal fats and has been implicated as a major cause of heart disease. Excess intake of fat leads to obesity, which puts extra strain on the heart and increases the concentration of fats in the blood. This causes deposition of fat on the inner lining of artery walls, causing a thickening of the walls and a consequent narrowing of the vessel – a condition called athero-sclerosis. The consequent roughening of the normally smooth walls and reduction in blood flow can be dangerous, especially if it occurs in the coronary arteries, and can lead to a heart attack.

Strongly implicated in the causes of this disease is a high level of CHOLESTEROL, a product of the breakdown of fat – especially saturated animal fats.

Unsaturated fats are not firm, so in the manufacture of margarine, refined vegetable oils are partly hydrogenated or saturated (the number of hydrogen atoms is increased) to increase the solidity. Often cultured milk, vitamins A and D, yellow vegetable dye and flavorings are added to make a product similar to butter. Some margarines have unsaturated vegetable oils whipped in to make them high in unsaturated fat.

Soaps

Fat is a basic ingredient in the manufacture of soap. The action of the hydroxide of sodium or potassium on fats produces *hydrolysis* or *saponification*, giving soap with glycerol as a by-product. Soaps made with potassium hydroxide, stearic, palmitic, and oleic fatty acid are highly soluble in water, producing soft soaps often sold as shaving creams.

The sodium-based soaps are harder and are sold as toilet soaps, usually with perfumes and germicides added. Some soaps used for scouring have an abrasive, such as pumice, added.

Glycerol, the by-product of this reaction, is used to make explosives (glycerol trinitrate), plastics, pharmaceuticals and antifreeze.

See also: Soap manufacture.

Feedback

Feedback is encountered in systems and devices which have an input into them, and an output resulting from this, part of which is taken back to the input to modify this in some way. When one part of a multistage process is coupled to an earlier stage in the same process, a *feedback loop* is established, which can be used to monitor the process and make any adjustment that may be necessary.

Although of great importance in engineering systems and electronic circuits in particular, feedback is a word commonly used in social contexts to indicate this process of *readjustment*. For example, a broadcasting company may obtain "feedback" from viewers or listeners on a program it has transmitted which will affect the nature of future broadcasts. Also, a teacher obtains feedback from the children on teaching methods employed and may modify the curriculum accordingly. Consequently, feedback is an essential feature of teaching machines where the pupils must be informed of their progress.

A simple feedback mechanism

A thermostatically controlled heater is a simple feedback system used to control the temperature of the room to a predetermined level set on the thermostat. The thermostat turns the heater off, the room then cools down, and at a given temperature the thermostat turns the heater back on again and so on, maintaining the room temperature within required limits.

The feedback loop in this situation is provided by the hot air traveling from the heater to the thermostat and the thermostat connection to the switch controlling the fuel input to the heater.

Left: The open loop (no feedback) amplifier at the top has a gain of 60. In an amplifier that has negative feedback (bottom left) one sixtieth of the output is subtracted from the input signal and the overall gain is only 30. If there is positive feedback (bottom right) the overall gain is 100.

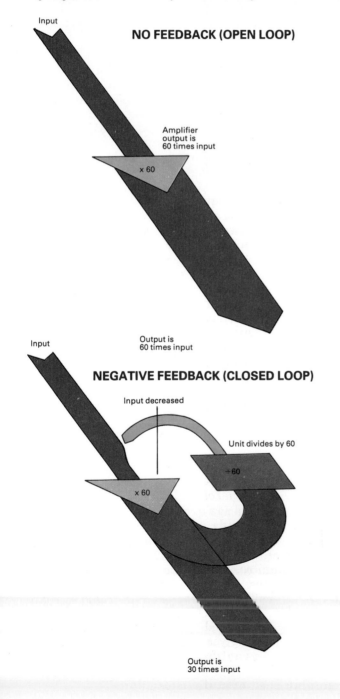

NO FEEDBACK (OPEN LOOP)

Input

Amplifier output is 60 times input

× 60

Input

Output is 60 times input

NEGATIVE FEEDBACK (CLOSED LOOP)

Input decreased

Unit divides by 60

÷60

× 60

Output is 30 times input

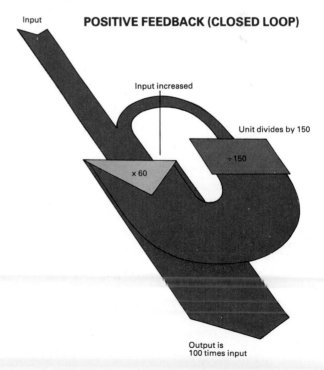

POSITIVE FEEDBACK (CLOSED LOOP)

Input

Input increased

Unit divides by 150

÷150

× 60

Output is 100 times input

Above: When playing live, bands rely on sound engineers to monitor inputs from microphones and instruments and modify the output through feedback.

Feedback in amplifiers

Feedback provides an extremely easy technique for controlling the gain (degree of amplification) of an electronic amplifier without altering any elements of the amplifier itself.

The amplifier alone has a certain fixed gain, called the *open loop gain* (open loop because there is no connection from the output to the input). This is the ratio of the output amplitude to the input amplitude which caused it.

When feedback is applied, a closed loop is created (from input to output back to input again). The feedback path includes an element which will allow only a certain fraction of the output amplitude back to the input. By controlling the size of this fraction, the feedback, and hence the closed loop gain of the system, can be controlled.

Positive and negative feedback

When the feedback signal is subtracted from the original input signal, the resulting type of feedback is negative. Negative feedback systems are inherently stable because any increase in the input signal is automatically counteracted by the returning (negative) feedback signal. The closed loop gain of such a system is therefore always less than the open loop (no feedback) gain of the amplifier. By controlling the fraction of output to reach the input, this closed loop gain can be controlled.

With positive feedback the feedback signal is added to the original input signal. In such systems, the gain can be increased beyond the open loop value. This is because the feedback signal increases the total input signal to the amplifier and a correspondingly larger output signal results.

There is, however, a limit to which the gain can be increased before the system becomes unstable. When the fraction of output reaching the input is the reciprocal of the open loop amplifier gain (as 1/6 is the reciprocal of 6), the system "lifts itself by its own bootstraps." The output careers out of control from the size of the feedback signal alone and no input signal is necessary. The system becomes unstable and oscillations occur – this is the principle of the OSCILLATOR.

Feedback theory played an important part in the development of radio for this very reason. Before 1913 high-frequency alternating voltages necessary for radio transmissions were impossible because ALTERNATORS could not be driven at the required high speeds. Positive feedback applied to the triode valve amplifier overcame this problem and enabled medium and short wave frequencies to be used.

Care is needed when designing feedback loops to insure that the feedback applied is of the desired polarity. If the signal being fed back is a regular waveform, a delay in the feedback path equal to half the period of the wave will be sufficient to turn negative feedback to positive and vice versa. Such delays may occur in circuits which contain capacitors or inductors. In some cases, the stray capacitances and inductances associated with the wires within the circuit may be enough to generate unwanted feedback.

When feedback is used for controlling large mechanical systems, it is normal to use varying amounts of three different types of feedback: proportional feedback is derived from the difference between the desired state of the output and its actual state, integral feedback is the time integral of proportional feedback, and differential feedback is derived from the rate of change of proportional feedback. While proportional control alone gives reasonably good performance, the addition of some differential feedback will increase the speed with which the output follows a change in input, and a little integral feedback will improve the accuracy of the output position when the input is static. Depending upon the relative magnitudes of these different types of feedback, the output may either approach its intended position ever more slowly, never actually achieving it (overdamped), move smoothly toward its intended position and remain there (critically damped) or overshoot and oscillate around its intended position (underdamped). When a large industrial process is being controlled, the oscillations may have a period of several days.

See also: Amplifier; Electronics; Flight simulator; Hi-Fi system; Oscillator.

Fermentation

Fermentation is a natural process used for thousands of years. Various alcoholic beverages are the most well-known of fermented products. Liquids containing fruit juice, when left open to the air, might discolor and give off bubbles of gas. Milk becomes sour if left without preservative treatment, and dead organic matter changes its composition after a time. Modern science uses the life processes of yeasts, bacteria and molds to produce chemicals and a large array of everyday products.

Microorganisms

Louis Pasteur, in 1860, was the first to understand fermentation. He showed that it was directly caused by the life processes of microorganisms, including bacteria, yeasts, and molds, which feed on organic materials. Yeast and bacteria are microscopic, unicellular organisms – yeasts are oval-shaped, bacteria more diverse in shape.

When conditions are right, yeast multiplies rapidly by a biochemical reaction which is used in the production of wine, beer, and bread. Bacteria multiply by binary fission – splitting apart – and are involved in a large number of different types of fermentation, including the production of cheese and vinegar. Molds are more complex microorganisms, consisting of multicellular filaments.

Organisms and energy

Usually, organisms obtain energy by breaking down sugar, in the presence of oxygen, into carbon dioxide and water. The reaction is:

$$C_6H_{12}O_6 + 6O_2 \rightarrow 6CO_2 + 6H_2O + energy$$
glucose   oxygen   carbon dioxide   water

Oxygen is essential for sugar to be broken down in this way. The process is called *aerobic respiration*.

If oxygen is not available, the organisms normally suffocate and die. In some cases, however, sugar can be broken down and energy obtained without oxygen. Instead of being broken down into carbon dioxide and water, other products are made. This kind of process without oxygen, *anaerobic respiration*, can occur in plants and animals.

One type of anaerobic respiration breaks down sugar and produces ethanol and carbon dioxide:

$$C_6H_{12}O_6 \rightarrow 2C_2H_5OH + 2CO_2 + energy$$
glucose   ethanol   carbon dioxide

This process is called *alcoholic fermentation* or, more usually, fermentation.

Another form of anaerobic respiration is the production of lactic acid in the muscles of someone taking strenuous exercise. The process supplies energy when insufficient oxygen is taken into the body. Sugar is broken down into lactic acid and energy is released.

Anaerobic respiration does not produce as much energy as aerobic respiration. For example, fermentation does not break down sugar completely – a considerable amount of energy remains locked up in the alcohol which may be released by burning. In aerobic respiration, the sugar is broken down completely. Though inefficient, anaerobic respiration is a way of surviving in oxygen-scarce environments.

Wine making

Fresh yeast looks like putty. In the wild, yeast grows on the surface of fruit and feeds on sugar. Essentially, sugar, water, and yeast are the only ingredients needed for making alcohol. In wine making, a process going back more than 4000 years,

Below: Two antibiotic crystallization units. After fermentation, liquid antibiotic is chemically purified then concentrated and crystallized. The unit is sterilized before each batch is processed.

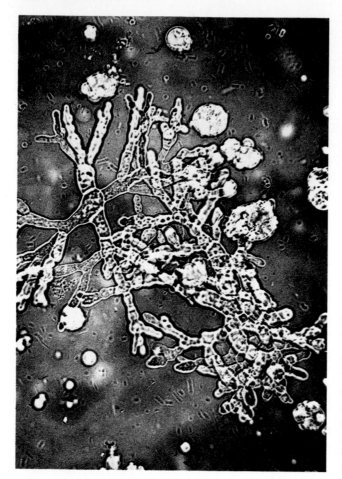

Above: The fermentation of *Penicillium chrysogenum* (under 750 × magnification) yields penicillin G – effective in the treatment of previously resistant strains of staphylococci.

the sugar is usually provided by the juice from crushed grapes. The juice contains sugar and wild yeast. As time progresses, the yeast feeds on the sugar, turning it gradually into alcohol.

The alcohol produced when making wine is always the same, but every wine has its own flavor, depending on the type of grape, and the exact conditions under which fermentation takes place. Wine making is not just restricted to grape-derived sugar. Wines can be made from berries, fruits or flowers.

Fermentation is usually carried out in a large, closed vessel fitted with a valve to allow carbon dioxide to escape and prevent air or airborne bacteria from entering and corrupting the wine. To keep the yeast active, the vessel is maintained at temperatures between 40° and 50° F (5° and 10° C).

Beer making

Beer making, or brewing, is a similar process to wine making. Various cereals are brewed with hops, which give the product a bitter taste. The important ingredient is barley. The grain is malted and *mashed* with water, then yeast is added, and the fermentation starts, converting the malt sugar into pure alcohol.

Alcohol is poisonous in large quantities. Beyond a certain concentration, it kills the yeast and stops the fermentation. For this reason, beer and wine never contain more than about 14 per cent alcohol. Spirits, such as whiskey, gin and vodka, need to be distilled after fermentation to achieve up to 60 per cent alcohol.

Bread making

Alcohol is not the only useful product when yeast breaks down sugar. The carbon dioxide given off is an important ingredient in baking. A mixture of sugar and yeast is added to the dough and the whole is left for about an hour in a warm place for the yeast to ferment the sugar. As the yeast cells multiply, they give off carbon dioxide gas, making the dough rise. During this period, alcohol is produced in the dough but it evaporates when the bread is baked.

Enzymes

Enzymes are the active constituents in fermentation (the word enzyme means "in yeast"). Usually, they can be extracted from the host microorganism, and work just as effectively when separated from the yeast cells. They can be thought of as biochemical catalysts – having a role in organic reactions similar to that of a catalyst in inorganic reactions. In common with inorganic catalysts, enzymes produce specific chemical changes, so the correct enzyme must be used in a particular fermentation reaction.

There are three main enzymes in yeast – *zymase, invertase,* and *maltase.* Zymase breaks up glucose and fructose; invertase breaks up cane sugar into invert sugar (a mixture of glucose and fructose); and maltase breaks up maltose into glucose.

The effect of fermentation is to break up the sugar into alcohol and carbon dioxide:

$$C_6H_{12}O_6 \rightarrow 2C_2H_6O + 2CO_2$$

glucose   alcohol   carbon dioxide

The fermentation processes that produce beer are a little more complex. The cereals used are barley, flaked rice, oats and corn. In Germany, wheat is used; and in China, rice and millet.

To make malt, barley is soaked in cold water, spread out on floors and turned regularly for five to eight days. During this period, germination starts, and the enzyme *diastase* is formed. When rootlets appear, growth is stopped by heating the barley gently. Diastase converts the starch in the malt into dextrin and maltose. At a later stage, yeast is introduced. Maltose is converted into glucose by maltase, and zymase starts the alcoholic fermentation. The enzyme used to make vinegar is the same one that causes beer and wine to go sour when exposed to the air. The process is called *acetic fermentation*, and is brought about by *Mycoderma aceti*. Generally, *Mycoderma aceti* finds its way from the air into weak alcoholic solutions. It lives on the nitrogenous matter in the solution and causes the alcohol to combine with oxygen from the air.

Fermentation and industry

By understanding how microorganisms function and how they interact with their environment, microbiologists can control fermentation for use in industrial processes. The production of lactic acid in 1880 was the begining of industrial fermentation. During World War I, a fermentation process was developed to produce acetone for use in the manufacture of cordite. Today, most antibiotics and several important chemicals are produced by fermentation.

Industrial alcohol is an important solvent used in the synthesis of other chemicals. It can be manufactured from many different sources, including wood, wood wastes, corn, and molasses. Fermentation has been found to be an uneconomic process, but it is subsidized by some governments, because of the products' importance.

Using fermentation to produce alcohol as motor fuel has been the subject of considerable controversy. Supporters argue that industrial alcohol, made from renewable biological materials, can be used to free the U.S. from dependence on foreign sources of petroleum. They also propose the use of surplus grain. Opponents claim that insufficient

Above: In huge fermentation vats like these in a British brewery, yeast, which breaks down the malt sugars and turns them to alcohol and carbon dioxide, forms a thick, creamy layer on the surface.

Below: Mash tuns at a Guinness brewery in Dublin, Ireland, in which grain mash is heated and agitated to activate enzymes and convert starch to sugar.

grain is available unless new technology is employed. Very pure alcohol will be needed if it is intended to mix with gasoline, requiring costly processing. Ethanol is used as an octane booster for gasoline, and, as petroleum prices rise, alcohol produced by fermentation will become economical.

Early fermentation processes yielded approximately two parts of butyl alcohol to one part of ethanol. This process used corn and the bacteria *Clostridium acetobutylicum*. Butyl alcohol became the most important product of the process, so there was a need to boost its production. Research produced new cultures which feed on molasses, producing three parts of butyl alcohol to one of acetone.

See also: Air; Alcohol; Beer and brewing; Catalyst; Enzyme; Winemaking; Yeast.

Vintage technology

New methods of cultivation, harvesting and fermentation are under constant research both in the traditional wine-making European countries — France, Germany and Italy — and in the new world of wine — the U.S., Australia and South Africa. All have one goal: to produce better, smoother wine, often at lower cost. But the traditions of wine and wine-making are preserved by a network of laws and the monopoly of the grape as its basic ingredient is jealously guarded.

Propagating wine-growing vines resembles propagating roses, by rooting cuttings and grafting, and keeping young plants in the nursery before transfer out to the vineyard.

Skillful pruning, beginning in midwinter, insures the optimum number of high-quality grapes. The partner of pruning, training, complements this process by producing the ideal vine for the latitude of the vineyard, the steepness of its hillside, and the microclimate involved.

In most vineyards, grapes are harvested in traditional fashion, by hand, so that defective fruit can be discarded at once. However, the more reliable crops of the French Midi and California, where the climate is predictable, can be picked mechanically.

To make red wine, the black grapes first go through a mechanical crusher. This practice has almost superseded the age-old one in which grapes were trod by foot. By breaking up the black grapes and their skins the crusher puts the yeasts in touch with the sugar. The grape pulp, or *must*, is then pumped into large vats. Traditionally made of wood, these may also be stainless steel vats or cement vats lined with epoxy resin, tile or glass. Sulfuring, the adding of sulfur dioxide, is then performed. This first absorbs oxygen from the must, then seals the must from the air by forming a coating over it. This coating kills the aerobic wild yeasts which will have already initiated the natural fermentation process, and also excludes the acetobacter.

After sulfuring the anaerobic wine yeasts continue the now-controlled fermentation alone. They go on working either until all the sugar is converted into equal parts of ethyl alcohol and carbon dioxide and other products such as glycerol and succinic acid, or until the alcohol content reaches 16 to 18 per cent. The wine yeasts then die.

The temperature for red wine fermentation, ideally between 77° F (25° C) and 86° F (30° C), must also be carefully controlled, for beyond these limits fermentation will stop. In general it is rarely necessary to raise the temperature of the wine by pumping it through heated radiators, except in very cold regions, as alcoholic fermentation generates considerable heat. To reduce the temperature must is pumped from the bottom of the vat to the top or, if greatly overheated, first pumped through cooling radiators.

Violent fermentation of the must lasts from one to four days, continuing at a less furious pace for up to four weeks. The *vin de goutte*, running wine, is

Above: Fungus growth keeps these 50-year-old vintage white wines at a constant cool temperature while they are maturing.

run off from the bottom of the vat when the required amount of color has been obtained from the skins. The mass of *marc*, skins left behind, is then removed through hatches at the bottom of the vats. The marc is then pressed to remove the concentrated juice known as *vin de presse*. Some of this *vin de presse* may be added to the *vin de goutte* by the wine-maker.

Pressing may be done by hydraulic screw press in which the screw rotates, crushing the marc between two endplates. Alternatively the more gentle cylindrical press can be used to produce the more delicate red wines. The cylindrical press employs a large rubber bag which is inflated by the rotation of the press to exert pressure on the grape skins. In large wineries, batteries of presses working like mincing machines are used.

Once the fermentation process is finished, the new red wine matures in wooden casks which are first carefully cleared of acetobacter. Steam is used to scald the casks, then sulfur is introduced.

The rest of the red wine-making is worked by time – from two to ten years of cellaring – during which the wine is carefully nursed. Carbon dioxide is allowed to escape but air and bacteria prevented from getting in by weights which seal the bungholes of the casks. Once a week, the casks are filled to the brim to combat the shrinkage which occurs with cooling and the proportion of wine which soaks into the sides of the cask. During the first three months of cellaring the wine is drawn from one cask into another to get rid of the sediment of dead yeasts. The process is known as *racking* and the wine is pushed into its new cask by pumping air into the old one, leaving the sediment undisturbed. Red wine is regularly racked once every three months, since its tartaric acid salts, coloring and tannins form a heavy deposit.

The cellaring period allows the wine to mature and also enables it to clear or *fall bright*. The tannin content does much to clear the wine, but not all. To complete the process, wines are *fined* (filtered) with a gelatine such as isinglass or with egg-white albumen before they are bottled.

White wines do not require quite the same complex process. The grapes are destalked, pressed and the juice at once separated to ferment without the skins. Lacking the tannin present in the making of red wines, white wines take between six months and one year to mature.

The halfway house of wine, the rosé, may start with grapes being processed in the same way as red wine. The juice is removed from the skins after a much shorter time – within one or two days – once the liquid has acquired the desired shade of

Above: Grape must – crushed grapes and the skins – is tested for sugar content in Germany's Gau Bickelheim laboratory.

pink. A much simpler method is to mix a little red with a large amount of white wine.

The latest technology in wine-making serves to adjust the natural process of production, frequently shortening the time involved. For example, in the carbon maceration of black grapes the fruit is placed in a carbon dioxide atmosphere which kills plant material in the grape skins and enables the colour to dissolve into the juice. This shortens the vinification process, making the wine ready for racking in 48 hours instead of four or five weeks.

Similarly, when confronted with a growing demand for white wine in 1979, California vinegrowers grafted white wine varieties onto existing red wine stocks.

Among the major pioneers of technology in wine-making are the Orlando vineyards in Australia. Orlando were the first to use pressure fermentation tanks in 1953 and first to try out cold fermentation.

However, with many types of wine nothing can take the place of time. Maturation is a never-ending process for the French Bordeaux and Burgundy wines, the German Hocks and the Italian Soave or Chianti. They go on maturing with their alcohol, fruit acids, pectins, tannin and other constituents of the original grape acting on one another until the moment they are consumed.

Fertilizer

Fertilizers are applied to the soil, or sometimes directly to the plant, to supply nutrients essential for plant growth. The fertilizer may replace these substances as they are used up by the plant, or may be applied to mineral-deficient soils to make them more productive. Many nutrient chemical elements such as copper, boron, manganese, zinc and silicon are only required in minute quantities (trace elements). Other elements are much more important, and are needed by the plant in large quantities. These major nutrients are nitrogen, phosphorus, potassium, magnesium, calcium, iron, and sulfur, all of which are readily removed from the soil by plants, particularly under conditions of intensive cultivation. Most fertilizers are intended to replace these important nutrient elements and their use has resulted in considerable increases in crop yields for farmers and gardeners.

Manures

Fertilizers are basically artificial mixtures of synthetic or naturally occurring substances. Manures, however, are composed of organic material, such as animal dung or various kinds of plant material. They often have the additional property of improving the texture of the soil, and giving it better water-retaining properties. Seaweed was formerly used for this purpose, and is still often used to help break down heavy clay soils.

Manures contain a high proportion of nitrogen, which is essential for plant growth. Atmospheric nitrogen is normally "fixed" and made available to the plant by bacteria present either in the soil, or sometimes in bacteria living in nodules on the plant roots. The plant can only use the nitrogen after it has been incorporated into a chemical compound. Most manures are produced by *composting*, which is a process designed to encourage bacterial breakdown of dung or plant material into a form readily used by the plant, and it is this process which results in the high nitrogen content of the manure.

Modern livestock farming techniques produce huge quantities of waste material, mostly as semiliquid cattle dung and poultry droppings, and the problem of disposal of this material without polluting the environment has led to renewed interest in large-scale manure production. Waste from intensive livestock operations is stored in artificial ponds or lagoons, and the solid matter

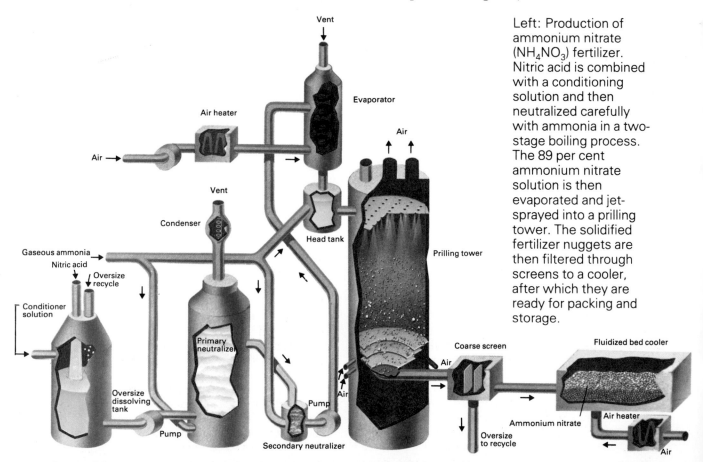

Left: Production of ammonium nitrate (NH_4NO_3) fertilizer. Nitric acid is combined with a conditioning solution and then neutralized carefully with ammonia in a two-stage boiling process. The 89 per cent ammonium nitrate solution is then evaporated and jet-sprayed into a prilling tower. The solidified fertilizer nuggets are then filtered through screens to a cooler, after which they are ready for packing and storage.

allowed to settle out while bacterial breakdown takes place. This nitrogen-rich solid material can then be used as agricultural manure. The sludge remaining after purification of sewage is also widely used as a manure, and is particularly in demand for horticultural purposes. A wide range of plant and animal waste products can be recycled in this way. Sometimes crops are grown especially to be plowed back into the soil as a manure. This is known as green manuring and is particularly effective when the manure crop is one that has nitrogen-fixing properties, such as the legumes. The classic example of this is the traditional European four-crop rotation method where clover is grown every fourth year for plowing in. A further advantage is that the green manure crop acts as ground cover and can help prevent erosion.

Manures contain variable amounts of nitrogen, and may be deficient in other major nutrient elements. As a good supply of nitrogen is important to stimulate the growth of leaves, manures have been largely replaced by fertilizers in the agricultural industry, allowing the quantities of nitrogen and the other elements to be carefully and scientifically controlled by the farmer.

Fertilizers

Ammonium sulfate, $(NH_4)_2SO_4$, and ammonium nitrate, NH_4NO_3, are typical nitrogenous fertilizers. They can be prepared in large quantities quite cheaply, and do not present problems in storage or applications to crops, in contrast to manures. Ammonium nitrate is produced by reacting ammonia gas with nitric acid. Ammonium sulfate can be made by reacting ammonia gas with sulfuric acid or by the reaction of calcium sulfate and a solution of ammonium carbonate.

Phosphorus is another very important element, often not present in the soil in sufficient quantities. It can be supplied in the form of bone meal, which is the powder remaining after crushing and grinding bones when other valuable chemicals have been extracted. Today phosphorus is usually applied in the form of superphosphate or triple superphosphate. Superphosphate is produced by treating naturally occurring rock phosphates with sulfuric acid, and contains about 20 per cent phosphorus. Triple superphosphate is manufactured by treating rock phosphate with phosphoric acid, and contains nearly 50 per cent phosphorus, while treatment with nitric acid produces nitrophosphates. The U.S. is a major producer of phosphate rocks with a production of over 35 million tons per year. Basic slag, a byproduct of steel manufacture, is another good source of phosphorus. Phosphorus is particularly important to establish good root growth, and to encourage early maturity and ripening.

Potassium is necessary to encourage good growth,

Above: Carrots on the left were fed with a liquid fertilizer containing nitrogen, phosphoric acid, potash, growth stimulants and trace elements (including iron, boron, zinc, copper, cobalt, and magnesium). Those on the right were untreated.

and helps produce resistance to disease. It is usually obtained from natural potassium sulfate, K_2SO_4, or potassium chloride, KCl.

Availability to the plant

Fertilizers and manures are used by plants at different rates. Manure cannot be used by the plant until it is broken down into soluble chemicals which can be absorbed by the roots. Some liquid fertilizers, such as those used for horticultural purposes, contain only soluble nutrients, and these are absorbed and used very rapidly. Foliar or leaf feeds are applied to the leaves rather than the soil, and are absorbed through the *stomata*, the tiny holes through which the plant breathes: 90 per cent of a foliar feed is absorbed and used by the plant, as against only 10 per cent when fertilizer is applied to the soil. Foliar feeds may also incorporate substances such as growth stimulants or trace elements which are not readily absorbed from the soil. Though relatively expensive, they are particularly useful in horticulture.

It is often desirable to use fertilizers in a slow-acting form, so that frequent reapplication can be avoided. Like manure, most organic fertilizers containing nitrogen break down slowly in the soil, and are available to the plant over a long period. Hoof and horn meal is a naturally occurring slow-acting nitrogenous fertilizer, and urea, $CO(NH_2)_2$, is a synthetic slow-acting source of nitrogen. It is often

used today in the form of urea condensates, that is, combined with formaldehyde, crotonaldehyde and isobutyraldehyde. These are, in fact, unstable resins (plastics) that break down in the soil, releasing urea.

If fertilizer is not applied properly, or at the wrong time, it can be washed or leached out of the soil before it can be used by the plants. Such leaching also contaminates the groundwater, and the U.S. Environmental Protection Agency has been promoting the use of contaminated groundwater for irrigation in order to reduce the amount of extra fertilizer required. An additional advantage is that this usage also helps to purify the groundwater.

Application to the soil or plant

Fertilizers are often compounded into granules or pellets, allowing a balanced mixture of essential nutrients to be supplied. Granules and pellets can be made which will break down at a predetermined rate to give a continuous supply of nutrients to the growing plants. Some crops need one or other of the major nutrients at different points in their growing cycles, so applications of fertilizers must be carefully timed to provide the correct nutrient balance when it is most needed.

Above: Liquid fertilizers and animal manures can be sprayed in a very similar manner. Here a cornfield is being treated with pesticides. The chemicals are sprayed out behind the tractor.

Fertilizers and manures can be applied to the soil or plant by a wide variety of means, ranging from simple hand application to machine spreading. Manures are often scattered on the soil and then plowed in. Sometimes liquefied waste from intensive livestock units is sprayed or pumped onto the pasture as a source of nitrogen. It has been suggested that this practice may spread livestock disease, and might also cause human infection with the organisms causing some types of food poisoning.

Most mineral or synthetic fertilizers are applied by topdressing, or mechanical spreading on the soil. This is sometimes carried out on a very large scale, such as the application of superphosphate over huge areas of inaccessible land in New Zealand by means of crop spraying aircraft. This resulted in a changed ecology throughout the area, making it suitable for sheep farming.

See also: Agricultural technology; Farming, intensive; Nitrogen; Phosphorus; Soil research.

Fiber, natural

Fibers are the raw material from which textiles are made for clothing, household, floor-covering and industrial uses. For convenience fibers are generally classified as being either natural or synthetic, that is, formed by chemical processes, usually involving extrusion of the fiber.

In 1983 the world production of natural fiber (excluding jute) was estimated at 16,954,000 tons, while production of synthetic fibers was 14,114,000 tons. Since 1971, world production of natural fibers has expanded only slightly, while production of synthetic fibers increased at a slightly faster rate. While synthetic fibers have relied on petrochemical expansion, the expansion of natural fiber production relies on developments in agriculture and competition for land utilization.

The use of natural fibers goes back to the Stone Age when flax and hemp were exploited. Eventually wool, silk and cotton fibers were utilized and were known to have been in use for several thousand years BC. In medieval times wool processing was a major occupation, but industrial processing, mainly of wool and cotton, dates from about 1750.

In modern times, all three natural kingdoms, animal, vegetable and mineral, supply textile fibers used in industry and at home.

Animal fibers

The hair of many mammals is potentially useful for textiles, but the principal fiber is sheep's wool. A wide range of breeds of sheep is used to provide fibers differing in fineness. The products vary greatly – from lambswool (baby wool), Merino (fine and soft), crossbred (medium wool for mixing and domestic fabrics) to upland and mountain types (coarse and wiry for carpets).

Wool, clipped or sheared from sheep, or *slipe wool* pulled from the skins of dead sheep, is not of constant quality from the same animal, and distinction is made between fine, coarse, outercoat and kemp hairs, which are thick white fibers. The sorting of wool, usually at the country of origin, is programed to separate the various grades of fiber to suit the intended use.

Each wool fiber consists of the protein *keratin*, and is built up of spindle-shaped cells. The main cell material or *cortex* is covered by a layer of thin overlapping scales (cuticle), which is visible under a microscope and gives wool its characteristic ability to felt or mat. Pigment streaks in colored wools are distributed in the cortex. Wool often has a *crimp*

Below: Flax being harvested traditionally in Belgium. Now this work is often mechanized. Flax was one of the first fibers cultivated by humans and is used to make linen and linseed oil.

wave which gives it springiness and warmth in finished products such as sweaters.

The principal producing areas are Australia, New Zealand, the U.S.S.R., South Africa and South America, although most countries with appreciable pasture produce wool.

Other animal fibers in use include goat hairs – chiefly mohair from the Angora goat, cashmere from the cashmere goat, and common goat – camel, llama, rabbit, horse, and cow.

Silk is of animal origin, and is the extrudate or spun thread of the silkworm *Bombyx mori*. Its natural function is to provide a cocoon around the worm, in which metamorphosis into a silk moth will take place.

Cultivated silk is produced under factory conditions, mulberry leaves being provided for the worms which, on hatching out of eggs, grow to several inches in length and contain twin sacs of *fibroin* or liquid silk protein. When it is ready to spin its cocoon, the worm attaches itself to a twig and extrudes the whole of its fibroin by muscular action through a small hole or spinneret, forming a cocoon of endless thread, several yards long. Since the

Below: Sisal is Kenya's third-largest export. It produces a coarse fiber that is used to make twine, rope and other rough fabrics in which great strength is needed. The fiber is inexpensive.

emerging moth would break open the cocoon and spoil the thread continuity, the cocoons are stifled with heat, steam or other gas. Next the cocoons are floated on water and *reeled*, several filaments being combined to form a fine and lustrous yarn. Japan and China are the main silk producing countries.

Wild silk, especially *tussah* and *anaphé*, is obtained from cocoons found in the open in Far Eastern countries. These cocoons are usually communal, and also broken, and cannot be reeled. The fiber is recovered by mechanical action and forms cut or staple fiber which is made into yarn by twisting. A typical fabric made from this source is *shantung*.

Vegetable fibers

These fibers are numerous and can be divided into seed hairs, bast (inner bark), leaf and fruit fibers.

Cotton is a seed hair, the plant producing numerous large seed pods from each of which radiate thousands of cotton fibers. These fiber clusters, or *bolls*, together with the seeds, are picked mechanically and processed in a cotton gin to separate the seeds from the hairs, which are subsequently spun into yarn.

The cotton fiber is a single cell, originally growing as a tube with a cell wall of nearly pure cellulose. When the boll bursts or opens, the fibers dry out and, when viewed under a microscope, resemble twisted flattened tubes. Cotton will spin into very

Above: A silk factory. First, the cocoons are smothered to kill the worms inside, then the silk must be carefully unwound and reeled. Cocoons of wild silk are broken and cannot be wound; instead, they are twisted to make a yarn.
Right: Wool. The use of animal and vegetable fibers is among the oldest of human accomplishments. Their production and distribution have always been of great importance to the world's economy.

fine yarns and is known for comfort and coolness. *Mercerization* is the treatment of cotton yarn or fabric with a concentrated sodium hydroxide solution, producing enhanced luster and smoothness.

Kapok is a seed hair, from a tree, and is used only for stuffing and (formerly) as a flotation material in lifejackets.

Bast fibers, or soft fibers, are recovered from the stems (phloem and cortex) of various dicotyledonous plants such as flax, jute, hemp, kenaf, surn and ramie (or China grass). Flax (synonymous with linen) is used for clothing, but the others are mainly used for industrial textiles and floor coverings. Jute is the most widely used member of this group.

The plant stems need *retting*, which is controlled rotting by soaking in water and allowing bacteria to attack the stems. The fibers are then separated mechanically by *scutching* (beating). The fibers are bundles of many overlapping cellulosic cells and are much thicker, stiffer and longer than cotton. The cells adhere by means of *lignin* cement.

Jute is grown in Bangladesh and India. Flax, formerly important in Northern Ireland, grows in the U.S.S.R., Poland, and Belgium.

Leaf fibers, or hard fibers, are similarly con-structed but are obtained from the leaf, stem or stalks of monocotyledonous plants. Mechanical separation only is necessary. Sisal and abaca (or Manila) are the main commercial examples. The main sources of sisal are Tanzania, Mexico and the Philippines, the last being the source of Manila, which is very rot resistant.

Fruit fibers are mainly represented by coir or coconut fiber, which is obtained from between the outer husk and shell of the coconut. The fiber is reddish brown, stiff and wiry, and is used for matting. Countries producing coir include India and Sri Lanka.

Mineral fiber

The term asbestos covers the fibrous mineral silicates, mined in Canada and South Africa. The rock form is often greenish in color, but after crushing the whitish soft fibers are obtained. These fibers are brittle and weak, but very resistant chemically and fireproof. Asbestos was therefore mainly used for insulation and protective clothing.

See also: Asbestos; Carpet manufacture; Clothing manufacture; Cotton; Textiles.

Fiber, synthetic

Synthetic fibers occupy a very important position in the world textile industry today, and they are produced in vast quantities by carefully controlled techniques. As with many other products, the story of synthetic fibers began with attempts to manufacture an alternative to a natural product – in this case silk. Synthetic fibers, however, could develop only as fast as the development of our understanding of the chemistry of fibers.

The first idea of manufacturing fibers was recorded as early as 1664, but it was some 200 years later before serious attempts were made to implement the idea. Then, in 1888, Sir Joseph Swan patented a process for making nitrocellulose filaments, subsequently converted to cellulose, for making not textiles but filaments for electric lamps. Within the next ten years Chardonnet had exhibited artificial silk fabrics at the Paris Exhibition, and the *cuprammonium* and *viscose* processes for making artificial silk from cellulose had been discovered. These fibers were later named *rayons*. The synthetic fiber industry had begun, and the production of viscose and cuprammonium rayons increased steadily. In the early 1920s another synthetic fiber – *cellulose acetate* – was introduced.

The next 20 years were a period of great activity in the study of polymer chemistry and this led to the possibility of producing completely synthetic fibers. *Nylon (polyamide)* was developed by Wallace H. Carothers in the U.S. in 1934 – in time to have an important use in World War II for making parachutes. *Polyester* fibers were first developed in Britain in 1941, and were launched commercially in the 1950s, as were several *acrylic* and *triacetate* fibers. At the same time fibers were being made from proteins derived from skimmed milk, ground nuts, maize or soya beans, but these could never compete with other new fibers.

Today, there are 20 generic types of fibers which are recognized in official lists, and this number is still increasing, but the consumer will usually only encounter some of these. For example, the important fiber groups for clothing, furnishings and household textiles are acetate, acrylic, nylon, polyester, rayon (viscose and *modal*) and triacetate. Other groups, such as *chlorofibers, elastofibers,* glass fiber, metallic yarns, *modacrylic, polyethylene*

Left: Syrupy orange viscose which is chemically ripened and then spun in a bath of chemicals to solidify the fibers into rayon. Most rayons are now produced by the viscose process, which was discovered in 1892. They are used for clothing, furnishing, carpets and industrial textiles. Below: Dry spinning acetate yarn by extrusion through a jet or spinneret into a heated airstream which evaporates the solvent, forming fibers.

and *polypropylene*, have important uses in the domestic market. Certain fibers such as glass fiber and other mineral fibers, carbon fiber and metal fibers and whiskers along with some plastics are also becoming of increasing importance for engineering applications.

The reason there are so many types of fiber is that each group possesses a different combination of physical characteristics, which determine their usefulness in the textile industry. A fiber can survive commercially only if it can contribute to the expanding performance and design requirements of the textile industry.

Fibers, both natural and synthetic, are composed of long chain polymers, with the molecules aligned roughly along the fiber length. Properties that a synthetic fiber polymer must have include good tensile strength and a high melting point, so that it will not disintegrate during ironing.

The normal manufacturing processes for synthetic fibers have certain features in common. The polymer is converted to liquid form by dissolving it in a solvent or by melting it, and the liquid is extruded through a spinneret (jet). The extruded filaments are solidified by precipitation or evaporation of a solvent, or simply by cooling. The filaments are then drawn (stretched) to cause the polymer molecules to lie more parallel to the fiber length, and this increases the strength of the fibers. Synthetic fibers are made in the form of continuous filament yarn or as staple fiber (short lengths). Staple fiber is twisted (or spun) together to give a spun yarn that is rougher to the touch but warmer than a single filament. Continuous filament yarns consist of several filaments spun together, and possibly textured, to give some of the characteristics of spun yarn. Fibers are also produced in a monofilament form where a single filament is thick enough for use on its own, such as for fishing line.

The characteristics of the fiber produced can be modified by alterations in the spinning process. For example the cross-section of the filament can be altered with a dog-bone shape giving a greater effective thickness and covering power than a simple round fiber. Inclusion of bubbles in the filament – before or after spinning – also gives the resulting yarn good covering power with a lot of bulk for the weight. Crimping the fiber, which can be carried out by mechanical or chemical means, gives it a wavy nature that improves the spinning qualities. Pigments can be added to the polymer to give a body color to the fiber.

Rayons
The word rayon has for years been used as a collective name for all synthetic cellulose fibers, but it is becoming more common to divide the rayons into two groups – viscose rayon (the standard rayon) and

Above: The glass-fiber body and deck of a plastic boat are molded individually. They are welded together with resin into one unit.

modal, modified from the standard rayon.

The terms viscose and cupro are taken from the names of manufacturing processes. The cuprammonium process, in which the cellulose is dissolved in cuprammonium hydroxide, is less common but is still used for producing special types of yarn. The viscose process, discovered in Britain in 1892, is the basis of the vast majority of rayon and modified rayon production today. The basic viscose process consists of several clearly defined stages.

The process starts by steeping wood pulp in a caustic soda solution, which converts the pulp into soda cellulose and at the same time extracts certain impurities from the wood pulp. The excess alkali is squeezed out and the swollen pulp is shredded into fine crumbs. After aging, the soda cellulose crumbs are treated with carbon disulphide to form the bright orange cellulose xanthate. This is dissolved in dilute caustic soda to give an orange-colored syrup known as *viscose*. After "ripening" to the right chemical state, the viscose is spun into a bath containing dilute sulfuric acid, sodium sulfate and zinc sulfate which causes the fiber to solidify at a rate which permits the yarn to be stretched.

The basic process has been modified in a number of ways to produce a range of fibers with quite different properties: high-tenacity viscose such as Tenasco and Tyrex for use in tire cords and other industrial purposes; crimped viscose for use in carpets, upholstery and clothing; high wet modulus (modal) for clothing and domestic textiles; and

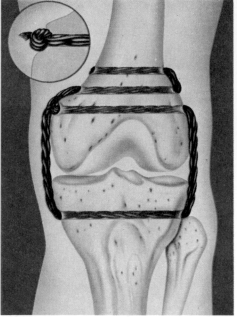

flame-resistant fibers. All rayons are absorbent and therefore give comfort in use. They are used extensively in all sections of the textile industry, sometimes alone, but often blended with other fibers.

Acetate and triacetate

Acetate and triacetate are the generic names for fibers made from cellulose acetate and cellulose triacetate respectively. Chemically, triacetate is a fully acetylated cellulose obtained by reacting cellulose from wood pulp with acetic acid in the presence of a catalyst, for example sulfuric acid.

Cellulose acetate is formed by diluting this solution of triacetate with water and allowing it to stand. Hydrolysis (chemical decomposition by the action of water) takes place slowly, some of the acetyl groups being replaced by the original hydroxyl groups of cellulose. This process is stopped by diluting with an excess of water when approximately one in six of the acetyl groups have been removed. The resulting product is known as cellulose acetate or simply acetate.

Fibers are produced by dissolving the polymer in an organic solvent, acetone for cellulose acetate, and methylene chloride for cellulose triacetate. The viscous solution is extruded not through a spinneret, as with rayon, but simply into a stream of hot air which causes the solvent to evaporate and the fibers to be formed. This is known as a dry spinning process as opposed to the wet spinning method used in the viscose process.

Cellulose triacetate fibers were produced as long ago as 1914, but the only known solvent at that time was chloroform – both expensive and dangerous. Also the fiber could not be dyed with the dyes then

Above left: Synthetic fiber being deposited into drums to dry at a factory in Spartanburg, South Carolina. Above: Torn and damaged ligaments in athletes can be painful and sometimes permanent injuries. They can be replaced, however, with carbon fiber substitutes.

available, and as a result the commercial introduction of this fiber was delayed for 40 years.

Cellulose acetate dissolved in acetone was used extensively in World War I for coating airplane wings and this left, at the end of the war, a capacity for producing cellulose acetate but no market for it. Cellulose acetate was easier to spin into fibers than the triacetate but was also very difficult to dye. Fortunately, in 1923 a completely new class of dye was discovered which would dye cellulose acetate fibers. This discovery was important not only for acetate in the 1920s but also in later years for the nylons, polyesters, acrylics and triacetates. Today acetate, sold under trade marks such as Estron, is noted for its silklike appearance and is used extensively in furnishings and clothing.

Triacetate, sold as Arnel and Tricel, is nearer to the fully synthetic fibers in that it can be durably pleated or set and has easy-care, quick-drying properties. Triacetate is used in dresses, knitwear, underwear, linings – and in household textiles such as bedspreads and bath mats.

Nylon (polyamide)

Nylon and polyamide are recognized alternative generic names for this group of fibers, nylon being use mainly in North America and Britain and polyamide in continental Europe. There are a number of nylons but they are all characterized by the presence of amide (—CO—NH—) linkages in the polymer chain. They are produced by the polymerization of an AMINO ACID or the corresponding *lactam* (a cyclic acid formed from amino acids) or by the copolymerization of a diamine with a diacid. The most common nylons are nylon 6 and nylon 6.6. Nylon 6 is produced by polymerizing caprolactam, and nylon 6.6 by copolymerizing hexamethylene diamine and adipic acid to form polyhexamethylene adipamide. Hexamethylene diamine, adipic acid and caprolactam are all produced by chemical reaction from oil refining products such as benzene or cyclohexane.

The numbers 6 and 6.6 refer to the number of carbon atoms in the chemical compound or compounds which are polymerized. A single number indicates a self-polymerized compound, while a double number denotes a combination of two compounds for the basic polymer link.

Nylon fibers are *melt spun*, that is, the polymer chips are melted in an inert atmosphere to prevent degradation, and the molten liquid is extruded through a spinneret. The filaments formed solidify by cooling. After spinning, they are cold drawn to several times their original length and this greatly increases the strength.

From the outset nylon was an astounding fiber. It was strong and resilient and hence hard-wearing. It was a thermoplastic fiber and could be heat set to give easy-care fabrics. Moreover the ability to be heat set enabled the development of crimped and stretch yarns widely used today in hosiery. Nylon took over most of the women's stocking market and, because of its high strength, finer gauge stockings became possible. Nylon today is used in almost every textile application, and has many industrial uses. It is manufactured in many countries under names such as Antron. Nylon fibers with trilobal (cloverleaf) cross-sections have been made to give the diffuse glitter seen in some fabrics, and antistatic nylons have also been introduced.

Polyesters

Polyester is the generic name for fibers made of polyethylene terephthalate – obtained from reacting ethylene glycol were terephthalic acid.

As with nylon, the raw materials are oil refining products, in this case paraxylene and ethylene, which are converted into terephthalic acid and ethylene glycol respectively. The fibers are produced by melt spinning and are drawn out subsequently to increase their strength.

Polyester fibers are produced in large quantities throughout the world under well-known brand names such as Dacron, Fortrel and Trevira. Polyester fabrics have set new standards of performance and easy care in suitings and dresses, shirts and household textiles. Suitings and trouserings have been mainly polyester-wool, polyester-cotton and polyester-rayon blends, but recently have included 100 per cent polyester knitted fabrics. Large quantities of knitted textured polyester and woven polyester-cotton are used in dresses. Polyester fibers are also used extensively in blends with rayon, acrylics, triacetate and nylons. Polyester, like nylon, has a wide range of industrial uses for example, as a component material.

Right: New methods of using plastics have led to a range of wonder materials with very different properties. For strength and flexibility, Glass Reinforced Plastic is superb, as this man shows.

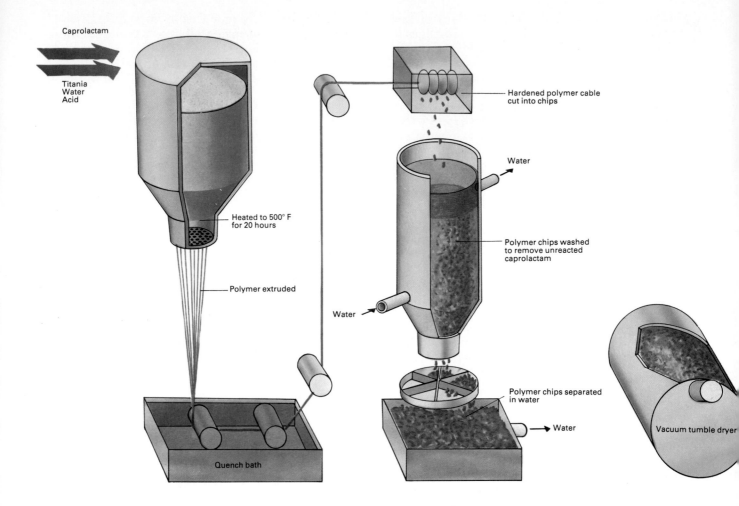

Above: Nylon 6 production. Caprolactam, titania and water are heated to 500° (260°C) for 20 hours until a polymer forms, which is then extruded into the quench bath for cooling and solidification. The polymer cable is cut into chips, washed and dried. In processing, these chips are reheated and extruded into a yarn.

Acrylics

Acrylic fibers are made from polyacrylonitrile obtained by polymerizing acrylonitrile. This chemical can be produced in a number of ways, one of which is to react propylene, another oil-refining product, with ammonia.

The fibers are often wet spun (extruded into an aqueous coagulating bath) but some are dry spun.

Acrylics are renowned for their softness combined with high performance and easy-care properties. Ingenious methods have been developed for producing high-bulk yarns to enhance the softness and warmth. The growth of this group of fibers has been very rapid and their brand names like Acrilan, Orlon, Courtelle and Dralon are well known. They are widely used in knitwear and knitted fabrics generally, pile fabrics and imitation furs, dresses, furnishings and carpets. They are increasingly used with nylon, polyester, wool or viscoses.

Modacrylics are fibers made from copolymers of acrylonitrile with vinyl or vinylidene chloride or both. They are very similar to acrylics in textile properties, but additionally can be flame resistant and are finding increasing uses where flame resistance is required.

Other groups of fibers

Although the remaining fiber groups are produced in smaller quantities, they nevertheless have important uses in the textile industry. The elasto-fibers (also referred to as elastomerics or spandex) have the generic names *elastane* and *elastodiene* in the International Standard list. They extend and recover like rubber, and are being widely used as power nets for support garments and swimwear under trade marks such as Spanzelle and Lycra.

Glass fiber is used for furnishing fabrics and for heat-resisting and fireproofing cloth as well as in optical fibers and as an industrial fiber for laminat-

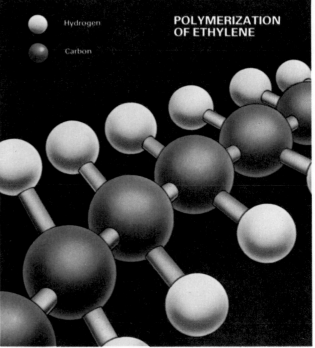

Right: Diagram of polymerization of ethylene – composed of two carbon and four hydrogen atoms. It yields polyethylene (polythene) whose molecular structure comprises thousands of carbon atoms linked in a long chain. Above: Polyethylene yarns are widely used for patio furniture, fishing nets, cordage and carpet backing. Similarly, carbon fiber's wide range of uses extends to parts for airplane engines.

ing with resins. Chlorofibers, made essentially from vinyl or vinylidene chloride, provide flameproof fabrics for furnishings and certain clothing. Metallic yarns such as Lurex are usually coated aluminum foil, to give a glitter effect.

Polyethylene yarns are well known as the material used for patio furniture and orange fishing nets. Polypropylene is used in cordage and carpet backing, and as slit film yarn in sacks and packaging materials.

Engineering fibers

Increasing awareness of the high strength offered by various types of fibers has led to their use for a wide range of engineering and industrial applications. Glass fiber is used for insulation purposes while aluminum silicate fibers have a similar use at higher temperatures of up to 2192° F (1200°C). The tensile strength offered by fibers is also being exploited by designers – one proposal for a bridge across the English Channel uses suspension cables of Parafil fiber to support the spans.

Carbon fiber is widely used for the production of composites and is produced as a type of rayon polymer which is heat treated to leave just the carbon skeleton of the original filament. This skeleton is then subjected to further mechanical and heat treatment to align the carbon crystals along the

fiber. Typically the fibers produced will be less than 0.0004 in. (0.01 mm) in diameter and have a tensile strength of up to 1250 tons per square inch.

Other materials such as aluminum, boron and silicon are also used to make fibers for use as reinforcement. In some cases the fibers take the form of whiskers which consist of a single crystal and offer much higher strengths than conventional fibers.

One of the most important engineering applications for fibers is in the production of composites such as glass fiber which consist of glass fibers held together by a resin matrix. Considerably higher strength can be obtained by using carbon or boron fibers with the resin matrix. Carbon and boron fibers can also be used to reinforce aluminum components (the aluminum is the matrix) and with a ceramic matrix. For higher working temperatures – around 750° F (400° C) – boron and molybdenum fibers have been used in a titanium matrix, while for even higher temperatures more exotic composites may be used.

These composites are finding an increasing number of uses ranging from sporting goods through car body panels to medical and aerospace applications. The advantages of such fiber reinforced composites vary according to the application but typically they offer high strength with low weight. In addition they can often be formed into complex shapes that would be very difficult and expensive to produce in other materials.

See also: Clothing manufacture; Fiber, natural; Polymer and polymerization.

Fiber optics

The principle which is used to illuminate fountains of water is used to advantage in fiber optics, thin glass fibers which can transmit light with little reduction in intensity.

The principle is that of *total internal reflection*. Every medium through which light can pass, such as water or air, has a certain *refractive index*, which determines the amount by which light is bent when it enters that medium. As the angle at which it strikes the medium decreases from the perpendicular, there comes a point where the light is bent so much at a surface that it is completely reflected back into the medium it originally came from – total internal reflection.

This takes place in columns of water and rods of glass, but light is usually lost through imperfections at the surfaces and through being scattered at the wrong angle for reflection to take place. In addition, the glass rod must be rigid.

These problems have been overcome by the development of fiber optics. An optical fiber consists of a cylindrical core of material (usually glass or plastic) clad with a material of lower refractive index. This prevents the light loss, while the thin-

ness of the fiber enables it to be very flexible. The light loss, or *attenuation*, in optical fibers can be very low. One experimental fiber reduces the light intensity by a factor of less than two over a distance of 0.6 mile (1 km), using infrared radiation, a figure comparable with electric cable attenuations.

Optical fiber manufacture

There are now several methods of manufacturing the fibers, the best known being the *rod and tube* method. In this process a rod of the core glass material, perhaps 1 in. (25 mm) in diameter, is assembled loosely inside a glass tube. This rod and tube assembly is then mounted vertically, with its bottom end in a small tubular furnace. At the appropriate temperature the rod and tube materials soften and, by pulling them downward, a clad fiber can be drawn onto a rotating drum.

Glass optical fibers are made with diameters varying from approximately 0.5 to 0.6 thousandths

Below: Making optical fibers. This method has concentric containers, the molten glass cooling as it leaves the furnace. The other is the rod-and-tube method. One type of fiber has a core, the refractive index of which increases outward; others have homogeneous inner fibers.

FIBER MANUFACTURE

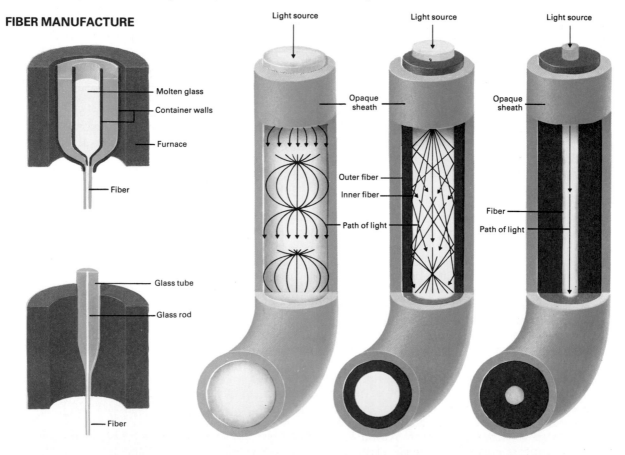

Molten glass
Container walls
Furnace
Fiber

Glass tube
Glass rod
Fiber

Light source
Opaque sheath
Outer fiber
Inner fiber
Path of light

Light source
Opaque sheath
Fiber
Path of light

Light source

of an inch (125 to 150 thousandths of a millimeter). (The diameter of a human hair is approximately 0.0002 in.) Plastic fibers can be made in larger diameters and still retain their flexibility. These fine fibers of glass or plastic are most often used as bundles containing at least several hundred individual fibers.

Applications

The uses of optical fibers cover an extremely wide range from the purely decorative to the most technical use as a communications medium. Most people are now familiar with the decorative lamp which uses a short thick bundle of fibers. The light, which can be varied in color, shows as a spread-out "tree" of points.

Most technical uses of fiber optics use either *incoherent* or *coherent* bundles of fibers. In an incoherent light guide there is no relationship between the arrangement of the individual fibers at the two ends of the bundle. Such a light guide can be made extremely flexible and provides a source of illumination to inaccessible places. When the fibers in a bundle are arranged so that they have the same relative position at each end of the bundle the light guide is known as coherent. In this case optical images can be transferred from one end to the other.

Perhaps the best-known application of optical fibers is in medical instrumentation. An incoherent light guide offers the best means of safely illuminating a part of the inside of the body. A coherent guide can then be used for observation or photography. The property of fiber light guides which is important in these cases is their flexibility. These kinds of instruments have various names, the common ones being bronchoscope, endoscope, gastroscope and cystoscope.

The most exciting modern application of optical fibers is in the new field of optical communications. Light transmitted down an optical fiber is equivalent to an electrical signal passing down a wire. The optical fiber, however, has a number of advantages over a wire. These advantages include a greater information-carrying capacity (a single fiber can carry many thousand telephone conversations) and a complete freedom from electrical interference. The extra requirement in this case for the optical fiber is the ability to transmit light with little attenuation over distances of hundreds of meters. In certain cases distances of many kilometers are involved. These demands have resulted in many advances in the methods of preparing the materials for making fibers and in the methods used for manufacturing them.

The speed with which light travels down an optical fiber varies with the initial direction of that light and its wavelength. For communication over long distances, single wavelength lasers are used

Above: Ultrapure glass strands capable of transmitting conversations at the speed of light – the heart of a fiber-optic cable seen installed at an exchange in Biarritz, France's guinea-pig city for advanced communications. Below: In Higashi Ikoma, Japan, residents are connected by a central computer to a two-way television system so they can send their own pictures back to the television studio. Such "wiring in" is made possible by optical fibers linking TV and telephone systems.

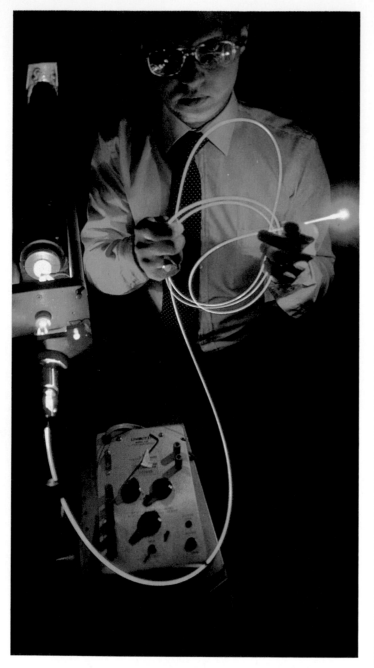

sive and can be used in many instances. A typical polymer cable will give an attenuation of 250 dB per mile compared with 8 dB per mile for a typical glass cable. It can also be safely bent around a curve five times tighter than glass although its tensile strength is less than that of glass. Terminations for polymer cables are fitted by cutting the cable and pushing a plug on the end, whereas glass cables normally require special cutting and polishing tools for preparing the cable which may then need to be glued into its plug.

Fiber optics are used in high-voltage equipment to provide electrical isolation. Control equipment may use large numbers of thyristors in series to switch high voltages. Each thyristor must be triggered at exactly the same time even though they are all at different voltages. This may be achieved by using fiber optics to carry the triggering signal.

See also: Electronics in medicine; Endoscope; Glass; Glass fiber; Laser and maser; Light; Optics; Optoelectronics; Plastics; Telephone.

• FACT FILE •

- When strain is applied to optical fibers, the way they transmit light is modified. Using this principle, researchers at Los Alamos National Laboratory in New Mexico have developed an optical-fiber sensor for detecting strains caused by earthquake activity. The laser-sourced sensor can detect strains as small as one part in ten billion.

- One medical use for optical fibers is the activating of light-sensitive porphyrins lodged in cancerous tissue. In this phototherapy, porphyrins are injected into the patient. Healthy tissue rejects them, but the cancerous tissue retains them. Optical fibers can be introduced into internal organs to transmit infrared light. Acting as a detonator, the infrared light triggers the porphyrins, which destroy the cancerous tissue.

- In October 1983, the Bell Telephone Laboratories succeeded in transmitting the equivalent of 6000 separate telephone signals through an optical-fiber channel 100 miles (161 km) in length. This was achieved without the use of repeaters to boost the signal, and involved sending 420 million *bits* per second down the line.

Above: Lasers produce a very fine but extremely powerful light. Here a laser is lighting a fiber optic instrument used for delicate surgery. Different lasers are used for special purposes. For example, surgeons can use argon lasers on the retina of the eye. These can cut and seal blood vessels without harming tissues.

and the refractive index is arranged to vary smoothly throughout the diameter of the cable rather than having a step change in order to minimize these problems. Glass is used for demanding applications but polymer cables are less expen-

Film

The first practical photographic system was introduced by L. J. M. Daguerre in 1839, followed in 1841 by Fox Talbot's *Calotype* process. Silver iodide was the light-sensitive material in both systems; in the former, as a thin film on a silver mirror, and in the latter, impregnated in the fibers of paper.

In both processes the *latent image* produced by exposure was invisible and was subsequently developed into a visible image. In Calotype, which was the ancestor of modern photography, the developed image was negative, that is the areas that had received the greatest exposure to light were darkest in the image. Positive images were obtained by printing the negative image onto another sensitized sheet of paper, the sharpness of the final image being limited by the fibrous texture of the paper.

In 1851 F. Scott Archer succeeded in precipitating silver iodide inside a layer of *collodion* coated on glass. Collodion is a viscous solution of nitrated cellulose, or guncotton, in a mixture of ether and alcohol. On drying it gave a tough transparent film from which excellent prints could be made. Since the developer solution could not penetrate the dry film, however, the plates had to be exposed and developed within minutes of being made, while the collodion was still wet, and this made the system rather unwieldy.

In 1871, Dr R. L. Maddox succeeded in replacing collodion with gelatin, a jelly made from the hides and bones of animals. Since dry gelatin reswells in water, these plates could be exposed dry and developed later. Separating the exposure from the manufacture of the plate and from its subsequent processing permitted the growth of a photographic manufacturing industry. The first true film was made in 1889 when the glass plate was replaced with a flexible plastic support, and this inaugurated the modern photographic era.

The basic photographic film consists of a thin layer of gelatin containing small crystals or *grains* of insoluble silver salts (silver chloride, bromide or iodide, which are collectively known as silver halides). This dispersion of particles in gelatin is called an *emulsion*. Individual grains vary in size from about one two millionth to one ten thousandth of an inch (about 0.01 to 3 μm) and the thickness of the layer is around a few thousandths of an inch.

Light decomposes silver halides into silver and the free halogen, which reacts with the surrounding gelatin or other materials in the layer. In a normal camera exposure the amount of silver formed is minute and can only be detected by development; it is therefore called a latent (or hidden) image.

A developer is a chemical that can convert silver

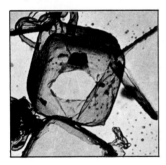 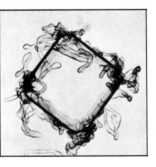

Left: Electron micrographs, at 15,000 times magnification, of silver halide grains in various stages of development. These grains will be fixed, removing the halide crystals to leave only the silver.

Below: The layout of a film coating track. At the far left, the rollers of the film loops are on runners so they will close up if the film base roller is stopped to splice on a new roll, without stopping the coating itself. After coating, the emulsion layer is very delicate so the fans used to chill and dry it must not be too fierce. The drying heat must not be so hot that infrared fogs the film.

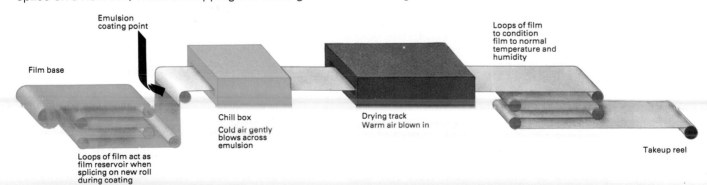

Emulsion coating point

Film base

Loops of film act as film reservoir when splicing on new roll during coating

Chill box
Cold air gently blows across emulsion

Drying track
Warm air blown in

Loops of film to condition film to normal temperature and humidity

Takeup reel

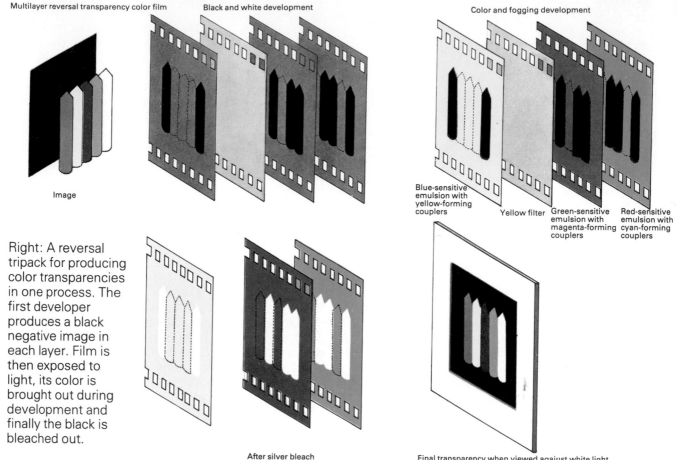

Multilayer reversal transparency color film

Black and white development

Color and fogging development

Image

Blue-sensitive emulsion with yellow-forming couplers

Yellow filter

Green-sensitive emulsion with magenta-forming couplers

Red-sensitive emulsion with cyan-forming couplers

Right: A reversal tripack for producing color transparencies in one process. The first developer produces a black negative image in each layer. Film is then exposed to light, its color is brought out during development and finally the black is bleached out.

After silver bleach

Final transparency when viewed against white light

halide into silver. The important requirement is that the reaction proceeds much more rapidly on grains possessing a small nucleus of latent image silver than on unexposed grains that do not. During normal development exposed grains are completely converted to silver, which is formed as a tangled mass of fibers and appears black, while unexposed grains remain relatively unaffected. Grains act as individual units, either being completely converted to silver or not at all, and intermediate tones are obtained by the complete conversion of some of the grains, and not by all of them being affected to only a certain extent. The grains in an emulsion are not all of the same sensitivity to light: there is a spread of sensitivity among them.

All photographic images are granular in structure, but the larger the individual grains composing them, the more obvious the granularity becomes on enlargement. Increasing the grain size decreases the ability of the film to record fine detail, but increases its sensitivity to light. The amount of silver that must be formed during exposure to render a grain developable does not depend on the size of the grain, while the amount of silver formed during development increases with grain size. The

amplification of the original effect of exposure that is obtained during development therefore increases with grain size too. The amplification that can be obtained on development is enormous. It has been estimated that the most sensitive photographic grains can be made developable by the formation of just four silver atoms. The development of a large photographic grain yields about one hundred thousand million atoms of silver: this is the basic reason for the high sensitivity of films.

To achieve good sensitivity to light and produce an acceptably grain-free image, grains of a range of sizes are used, the small grains providing fine detail while the larger ones provide sensitivity.

After development the unchanged silver halide grains are dissolved, or *fixed*, usually in a solution of sodium or ammonium thiosulfate. Unwanted chemicals are washed out and the film dried.

In converting silver halide into silver a corresponding amount of developer is also changed, or oxidized. In black-and-white photography the silver image is wanted and the oxidized developer is eliminated by fixing and washing, but in color photography the silver image is eliminated while the oxidized developer is used to make the dye image.

Right: Making a color print using negative film and paper. The first development stage produces colors that are complementary to the originals. The mask is partly formed in two layers, with the result appearing orange. Printing out on paper again produces complementary colors, that should be the same as the colors of the original subject.

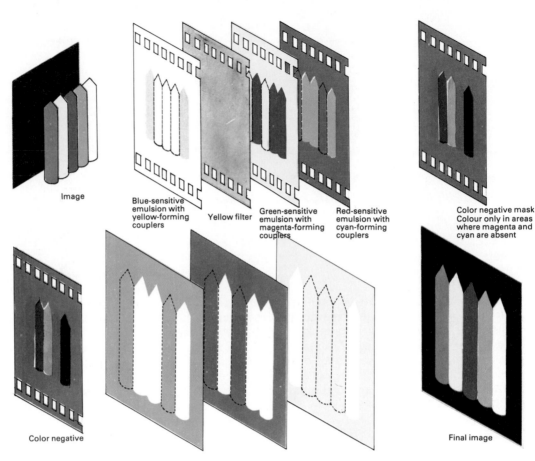

Image

Blue-sensitive emulsion with yellow-forming couplers

Yellow filter

Green-sensitive emulsion with magenta-forming couplers

Red-sensitive emulsion with cyan-forming couplers

Color negative mask Colour only in areas where magenta and cyan are absent

Color negative

Final image

Some black-and-white films designed to have a very wide exposure latitude (range of exposure which will provide a satisfactory image) also replace the silver image with a dye. A range of different black-and-white films and developers are available with differing characteristics, such as high sensitivity, low contrast, high contrast, and fine grain.

Silver chloride grains are only sensitive to ultraviolet radiation. Although the sensitivity of silver bromide and even more of silver iodide extends into the visible spectrum, it is still restricted to the blue end. Dyes absorb light of different colors and some of them are capable of transferring the energy of the light they absorb to photographic grains. These dyes are used to sensitize the grains to light of any required color, throughout the visible spectrum and even to near infrared radiation.

Early black-and-white films were not dye sensitized, and were only sensitive to blue light, resulting in odd tone renderings for colors. Eventually it became possible to sensitize the emulsions to green light as well, giving *orthochromatic* films; fully color-sensitive black-and-white films are called *panchromatic*.

The sensitivity or *speed* of a film is rated so that the less sensitive, *slow* films have low numbers while the more sensitive, *fast* films have high numbers. Two systems are in use, ASA and DIN. On the ASA system, a doubling in sensitivity doubles the speed rating; on the DIN system, it adds 3 to the speed number.

The ASA and DIN ratings are now commonly combined into a rating called ISO. Thus a medium speed black-and-white film has a speed of 100 ASA, 21 DIN or 100/21 ISO while a film of twice that speed, such as a color negative film, is 200 ASA, 24 DIN or 200/24 ISO. Many 35 mm films include their designated speed in the form of a series of black and silver bars on the outside of the cassette called DX coding. This code can be read by sprung metal pins in some cameras and used for automatically setting the camera exposure system. DX coded films also carry coded marks along the edge of the film strip. After development, these may be used by the film processor to control automatic printing machinery.

Color films

The eye contains different receptors which are mainly sensitive to red, green and blue light respec-

Above: After the material (in this case paper) has been coated, it is slit, as shown here, or cut into sheets. The emulsion has turned pink from the lighting for the photograph.

tively. Light of other colors is perceived by the relative extent to which these three kinds of receptors are excited. Modern color films contain three photographic layers, one sensitized to blue light, one to green and one to red, coated on top of one another in a *tripack*. Because sensitizing a layer to green or red light does not destroy its natural sensitivity to blue light, the blue-sensitive layer is put on top. A yellow filter layer, which absorbs blue light, is interposed between it and the underlying layers to prevent blue light reaching them.

Besides photographic grains, the layers of a color film may contain color *couplers*. These are compounds which react with the oxidized form of certain developers to form dyes. The formation of silver during development is accompanied by the formation of a corresponding amount of oxidized developer and this, in turn, reacts with the coupler to form a dye image. The coupler in the red-sensitive layer is chosen so that the dye formed absorbs red light and therefore appears as the complementary blue-green (cyan) color. Similarly a red-blue (magenta) dye is formed in the green-sensitive layer and a yellow dye in the blue-sensitive. Because the couplers must remain in their appropriate layer, they are either given bulky molecules or are dissolved in tiny oil droplets to prevent them from diffusing through the gelatin and cause blurring.

Development is followed by a treatment which removes both the silver component of the image and the undeveloped silver halide in a *bleach-fix* bath. The yellow filter layer, which usually consists of very tiny silver particles, is also eliminated in this step. Then the film is washed. The image obtained is not only negative in terms of tone but also complementary in color to the original object. Printing this negative image onto a similar tripack, either on film or paper, gives a positive image in the correct colors.

Direct positive films, which are used for amateur movie film and color transparencies, are constructed similarly but processed differently. They are first developed in a developer which does not react with the couplers, to give a negative silver image but no dye. Then the undeveloped grains are made developable by exposing them to light or incorporating a suitable *fogging* compound in a color developer. This process yields a direct positive dye image and the unwanted developed silver is removed as before.

The dyes used in most films are impure colors. This results in imperfect rendering of some colors. In negative films, this effect is avoided by tinting each of the color couplers with the same imperfections as those which will be present in the dyes which they form. This masking means that the imperfections are evenly distributed throughout all colors and can be corrected by using suitable color filters during printing. Corrective color masks are the cause of the overall orange appearance of most color negatives.

Color masking cannot be used with direct positive films. For very high quality work, special films which incorporate very pure dyes are used. During development these dyes are destroyed where they are not needed to give a long-lasting, accurately colored image.

Manufacture

Photographic *emulsions* are prepared by mixing solutions of silver nitrate and of sodium or potassium halides in a hot solution of gelatin. The characteristics of the resulting emulsion are determined by the choice of halides (chloride, bromide or iodide or mixtures of them), by the rate of mixing and the presence of solvents for silver halide. Slow mixing and the presence of solvents, which help the smaller grains to redissolve and reprecipitate on larger grains, encourage the formation of larger, more sensitive grains. The gelatin controls the growth of the grains and prevents disorderly clumping.

The unwanted sodium or potassium nitrate formed in the reaction is removed by coagulating the gelatin into lumps or curds. The coagulated gelatin settles out, carrying the photographic grains with it, while the unwanted nitrates remain dissolved in the water, which is discarded. The gelatin and the photographic grains are then redispersed in clean water.

The grains are made more sensitive by adding a small quantity of sulfur or gold compounds to the emulsion and heating it to about 120° to 130° F (50° to 55° C). This leads to the formation of small quantities of silver and gold sulfides on the surface of the grains and increases their sensitivity. Then the sen-

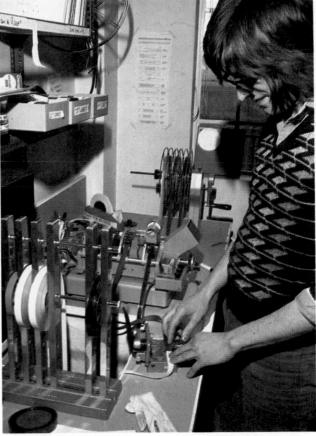

Above: The sound track of a film is recorded on tape and transferred to a magnetic strip along the edge of the film. Music and other sound effects are spliced together (right) in the correct sequence and linked up with the sound track. Some films have the sound track recorded optically as variations of light and dark areas.

sitizing dyes, color couplers and compounds to improve the stability of the emulsion, to reduce fog and to harden the gelatin are added. The emulsion is then ready for coating.

Film base comes in large rolls, several feet wide and several hundreds of feet long. Different kinds of plastic are used but they must be nonflammable, retain their flexibility over long periods and be impervious to water, so that their dimensions do not change in processing solutions. They must also be cast with a very smooth surface. Photographic papers are coated with *baryta*, a layer of a white pigment in gelatin, to improve whiteness in the highlights, before being coated with emulsion.

For coating, the molten emulsion is pumped through an accurately machined slot and allowed to flow evenly onto the film base as it is drawn past the coating point. The film is then chilled to set the gelatin. This is important, since the set gel cannot flow and the layer remains even while drying. The dried, coated film is then wound up onto a roll. X-ray films are coated exceptionally thickly to absorb sufficient radiation, and are usually coated on both sides of the base; color films receive several layers on top of one another and most film materials are given a dark *antihalation* layer on the back to absorb light that passes through the film and keep it from being reflected back onto it, producing haloes around bright lights. The film is then cut into strips of the required width, perforated (if necessary), chopped into lengths and wound on spools,

before being packed in cassettes or being wrapped.

Besides films for normal use, special films are made to record X rays or the tracks of nuclear particles. Films with particularly fine grains are made for microphotography and for use in the manufacture of integrated electronic circuits. Films are specially sensitized for making the very long exposures needed in astronomy and false color films sensitive to infrared radiation are widely used in aerial reconnaissance.

In photographic terms the range of photographic papers is more limited. They are, however, manufactured with a range of contrasts and finishes.

The bases of most color printing papers and some black-and-white printing papers are coated on both sides with a layer of plastic which makes them impervious to chemical solutions and allows quicker processing since less washing is required. Such papers are termed resin coated. Traditional papers, termed fiber based, are completely porous and therefore require more thorough washing. The lack of a plastic backing also makes them more difficult to dry flat and more liable to be damaged by rough handling when wet.

See also: Camera; Exposure meter; Light; Photographic processing; Plastics; X ray.

Filter, optical

Optical filters are used to eliminate or reduce the intensities of certain color components in a beam of light. This is achieved by either passing the beam through, or reflecting it off, a material with the required color properties. Optical filters are used in a variety of applications including photography, color TV, stage lighting, protective eyeglasses (goggles) and spectroscopy.

Just as electrical filters reject certain frequencies, so optical filters absorb or reflect particular wavelengths or frequencies of light. The spectrum of ELECTROMAGNETIC RADIATION ranges from red light through the colors of the rainbow to violet. Together, these colors produce white light. If a band of color is removed, the remaining colors will predominate: white light passed through a filter which absorbs blue and green light will then contain yellow, orange and red light, and will have an overall orange color.

Absorption filters

The most common type of filter uses the absorptive properties of various dyes and minerals. These may be mixed with the transparent filter materials, which may be glass, plastic, gelatin or cellulose ester; or they may be coated onto a surface of material and then protected by sandwiching them in glass.

Below: These pictures show the effects of various gelatin dye filters on a spectrum projected onto a screen. The shots show the layout – the filter affects the lower half of the spectrum only. Here the color of the filter itself can be seen in the lower left-hand corner – magenta at left, cyan (peacock blue) at right. The magenta filter cuts out green and yellow light completely, while the cyan cuts out yellow, red and some violet light.

Above: Using white light, the transmitted ray appears blue while the reflected ray appears yellowish. In color-television cameras the image has to be split up into three colored components, red, blue and green.

When light encounters molecules of certain materials, the photons or packs of light energy of particular wavelengths may force these molecules to resonate. The light is re-emitted, but at a longer (infrared) wavelength, at which the photons have less energy. This means that the beam of light has lost energy at one wavelength, or in practice a band of wavelengths. Some wavelengths are more easily absorbed than others: it is particularly difficult, for example, to make a filter which will absorb only red light, transmitting blue. Blue-colored filters tend to absorb a little of all colors.

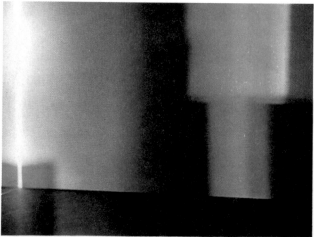

The gelatin is extruded through a thin slit onto a very flat moving glass plate as a layer about 1/25 in. (1 mm) thick. This is then chilled to set the gelatin so that it can be peeled off the plate for drying. On drying it shrinks to a thickness of only about 0.004 in. (0.1 mm).

If a more robust filter is required, the filter materials may be bonded onto optically polished glass and then sandwiched by another piece of the same type of glass. Alternatively an extremely hard, durable resin may be used to make the filter, the surface of which will absorb the coloured dye.

Dichroic filters

With normal filters the light that is absorbed is lost, that is, unusable. But there is another type, known as *dichroic* or *interference* filters, where the non-transmitted light is reflected. Their name (meaning two colors) comes from the fact that the light reflected off them contains every color that was present in the original (incident) beam except the color that is transmitted through the filter. Furthermore, by controlling their fabrication to within very fine limits it is possible to allow a very narrow band of wavelengths through the filter. Dichroic filters are used, for example, in color television cameras where the reflected part of the beam can be used rather than wasted.

Dichroic filters are also known as *dielectric* filters because of the particular electrical properties of the materials – usually ceramic materials – used in their construction. These materials are laid down in very thin and even layers on a glass base (called the *substrate*) by evaporating them in a vacuum chamber to make the filter.

The action of these filters depends on the interference effects of light waves reflecting from the surfaces of the various layers. The number of layers, and the thickness of each, determines the final properties of the filter. Generally, the layers are less than 0.00002 in. (0.0005 mm) thick; three such layers will transmit about half the visible spectrum, about 1800 angstroms, but with 17 layers the transmitted bandwith is narrowed down to about 10 angstroms. Extremely narrow transmission bandwidth filters (less than 1 angstrom) have two sets of narrow layers separated by one thick one. (1 angstrom unit equals 10^{-7}mm.)

Filters tend to be given descriptions which vary with their purpose. The filter used to make red traffic lights is always called red, though it actually filters out blue, green and yellow light. On the other hand, to make their photographs clearer, photographers often use filters which they call ultraviolet, to absorb the blue and ultraviolet wavelengths which tend to produce a haze on photographs of distant scenes, though these filters may actually be colored light yellow.

Materials

The types of filters which can be made depend on the materials. The earliest filters were glass, but now gelatin filters are more popular.

Colored glass can be made by simply adding various metal oxides to the molten glass. Iron oxide, for example, produces a green color and is present anyway in poor quality window glass. Cobalt oxide produces blue coloring, manganese purple, and so on. The range of oxides available is limited, but some of them have unique properties, such as the *didymium* glass, which absorbs the yellow light of sodium streetlights while letting most other colors through.

The dyeing industry has produced a vast range of dyes, many of which can be mixed with gelatin, the material which is used to prepare the emulsion of photographic film. This is made very much thicker than photographic emulsions, since it will be used as a thin gelatin sheet and not coated on a flexible backing.

See also: Electromagnetic radiation; Glass; Light; Optics; Television techniques.

Below: A starburst filter – a type of diffraction grating – usually has a rotating mount which can be turned to position the spikes of the star to help the photographer compose the picture.

Firefighting

Ever since the discovery of fire, humans have realized the importance of being able to control it, as an uncontrolled fire can cause widespread devastation. Throughout history there are examples of cities being devastated by fire, for example Rome in AD 64, London in 1666 and Chicago in 1871.

Among the first organized firefighting forces were those established throughout the Roman Empire. Slaves were employed as firefighters and were stationed on the walls and at the gates of cities. This organization, known as the *Familia Publica*, was later replaced by the corps of *Vigiles*, who also had other duties which included the capture of runaway slaves and policing the city at night. After being in existence for more than 300 years the Vigiles were put on an equal footing with the Imperial troops, although they were not classed as soldiers. They were also responsible for preventing fires, and they had powers to punish offenders.

After the decline of the Roman Empire, the countries which had been under Roman rule were left to make their own arrangements for fighting fires, and initially they did not do very well. Hundreds of years passed before any effective organizations evolved. In Europe it became customary to ring a bell at night to order people to put out their fires and candles. This was known as the curfew, from the French *couvre-feu* (cover-fire).

Small towns and villages usually kept a supply of buckets, hooks and sometimes ladders available for fighting fires – basic items of equipment which will be found on even the most modern fire engine. As towns grew larger and more industrialized the basic equipment supplied by the local parish authorities was soon found to be inadequate. Portable pumps were introduced, usually operated by volunteers, but there were seldom any paid, responsible officials to look after the equipment. Contemporary records often mention instances of broken pumps and missing buckets, inadequacies which were never found until it was too late.

Insurance companies

In the eighteenth century the new business of insurance was gaining ground, and the early fire insurance companies took over the responsibility of providing professional firefighters and building fire stations. They also provided a fire mark or badge to identify the premises that they insured. When a fire occurred several different fire engines would be summoned, and the first thing they did when they arrived was to look for the fire mark of their own company. If the mark was there, the fire would be dealt with; if not, the firefighters would often return to their stations. It was inevitable that petty rivalries would occur. Sometimes the firefighters from one company would stand and watch other putting out the fire, often jeering or even obstructing them. The hand-operated pumps were operated by volunteers who were often paid with beer, and fights were not uncommon.

Eventually the responsibility for firefighting was taken over by the local authorities. In Britain, there were about 2000 separate fire brigades in operation when they were nationalized following the outbreak of World War II, and these were returned to local authority control in 1948.

Fire appliances

Today's fire appliances perform the same basic function as those used by the insurance company fire

brigades in the early nineteenth century, that is, they carry the firefighters, equipment and pumps to the scene of the fire as quickly as possible.

The equipment used by firefighters is more complicated these days, and is as diverse as the types of fires they have to attend. In fact there is so much equipment that fire appliances have become specialized units, and broadly fall into several groups, each group having a particular role to play.

A firefighter's first responsibility is rescue, and the firefighting must take second place to efforts to save life. The first group of appliances is therefore aimed primarily at life-saving and secondly at firefighting.

Typical of this type of appliance is the *pump escape*, widely used in Britain and many other countries. These machines have a powerful engine, usually with automatic transmission, and a pump capable of delivering up to 1000 gal (4550 l) of water per minute. They carry a wheeled ladder which extends to about 50 ft (15 m), equipment to break in through doors and windows, rescue lines to lower people from windows, lights to help penetrate the gloom of a smoke-laden building, breathing apparatus, hoses and jets (hose nozzles). With its crew of five, this appliance is the basic rescue and firefighting unit of the modern fire service. Basically similar machines used in the U.S. are known as *triple-combination pumpers*. The most modern variant of the pump escape is the *pump hydraulic platform*, which is similar in most respects except that it carries a 50 ft (15 m) hydraulically powered

Left: Firefighting vehicles at work. In the foreground is a tanker holding foam for the hoses. In front of it is the aerial ladder truck whose ladder can extend up to 100 ft (30 m). Many different types of equipment are necessary to control a large fire.

Below: A Pathfinder heavy duty airport crash truck in action with a burning aircraft. Weighing 37 tons, it can accelerate to 50 mph (80 km/h) in 50 seconds.

Above: Firefighting in 1673. This illustration is from a German book and shows a pump, mounted on a sleigh, which is hand-operated and filled from buckets of water.

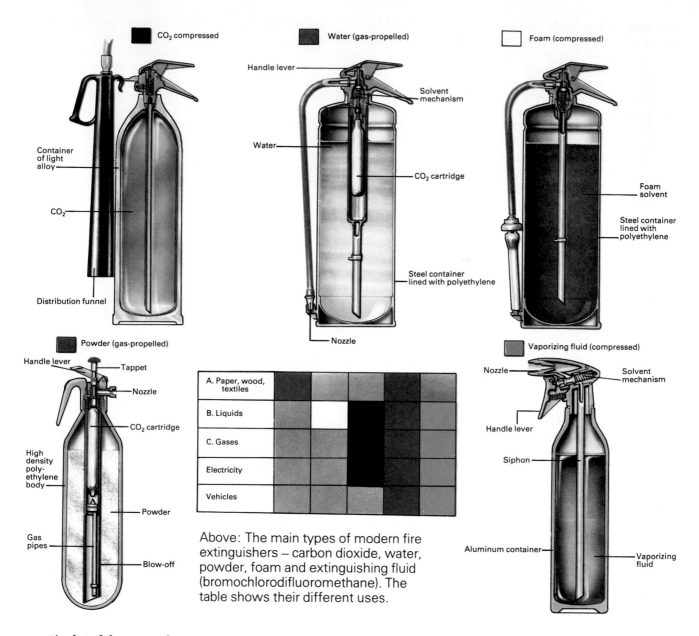

CO₂ compressed

Water (gas-propelled)

Foam (compressed)

Handle lever

Solvent mechanism

Container of light alloy

Water

CO₂

CO₂ cartridge

Foam solvent

Distribution funnel

Steel container lined with polyethylene

Steel container lined with polyethylene

Nozzle

Powder (gas-propelled)

Vaporizing fluid (compressed)

Handle lever

Tappet

Nozzle

Solvent mechanism

Nozzle

CO₂ cartridge

Handle lever

High density poly-ethylene body

Siphon

Powder

Gas pipes

Blow-off

Aluminum container

Vaporizing fluid

A. Paper, wood, textiles					
B. Liquids					
C. Gases					
Electricity					
Vehicles					

Above: The main types of modern fire extinguishers – carbon dioxide, water, powder, foam and extinguishing fluid (bromochlorodifluoromethane). The table shows their different uses.

articulated boom and rescue cage in place of the wheeled escape ladder.

The second major category of appliance used in Britain is the *water tender*, which supports the rescue appliance and carries about 400 imperial gallons (1820 l) of water in addition to the basic firefighting equipment of breathing apparatus, hoses, jets and ladders. In addition to the main pump there is a portable pump so that the firefighters can take advantage of any natural water supply such as a stream or pond which is inaccessible to the tender.

Other basic appliances in use in the U.S. are the *service ladder* truck, which carries a range of ladders; the *combination pumper-ladder* truck (the quad); the *aerial ladder* truck, which carries a powered ladder (similar to the British *turntable ladder*) as well as a range of ordinary ladders; and a combined pumper and aerial ladder unit known as the *quintuple* truck.

The turntable ladder appliance enables the firefighter to work at a greater height. It carries a ladder which will extend to 100 ft (30 m) or more, mounted on a turret that enables it to rotate through 360°. All the ladder movements are hydraulically operated and powered by the engine of the appliance, which is made into a stable working base by the use of axle locks to prevent the road springs from flexing, and outrigger jacks to steady the vehicle. The firefighter at the top of the ladder has a telephone to enable communication with the driver below who is operating the controls. This appliance is used for rescue purposes and to put large quantities of water onto a fire at a high level. When in use on a hill or on a road with a steep camber, the vertical plane of the ladder is maintained automatically throughout the rotation movement, thus assisting stability.

Many fire departments are now using the larger versions of the hydraulic platform (known as *snorkels*) for fighting fires in tall buildings. First introduced in Chicago in 1958, they can provide a stable working platform for several firefighters at heights of around 70 ft (21 m), with a pivoted jet mounted on the platform supplied with water through built-in pipes.

Support vehicles

All the appliances mentioned so far are used for the actual fighting of fire, but at any large fire the firefighting team requires additional facilities that are provided by a range of support vehicles. Hose-laying vehicles are used when long lines of hose are needed to provide large amounts of water. When special equipment, such as breathing apparatus, light, power tools, protective clothing, and radiation or gas monitoring meters, are required, they are supplied by an *emergency tender*.

The effects of smoke and water on the parts of a building not affected by the fire must be considered, and so a *damage control unit* or *salvage tender* is used, which carries equipment to help minimize any hazards and takes effective steps to control damage.

At a large fire, communications will be centered on a *control unit*, which enables firefighters to maintain radio contact with their headquarters, other appliances and personnel, fireboats, rescue vessels and helicopters.

Boats and aircraft

Fires do not always occur in places accessible to wheeled vehicles; for example, at ports one of the hazards to be faced is that of fire on board ship. To deal with this situation firefighting is carried out

Above: Fireboats at a dock fire in warehouses in London. This picture is not of a real fire; it was shot during the making of a film.

from *fireboats*, powerful boats fitted with pumps and all the equipment carried on land-based appliances. For work in harbors, docks and river estuaries, these boats are basically powerful motor launches, but larger versions based on tugboat designs are capable of dealing with fires on ships at sea.

An increasing use is being made of helicopters for survey, control, and rescue work in large cities. In Britain, the fire brigades often work in conjunction with Royal Air Force rescue helicopters, but in many large cities such as Tokyo and Los Angeles, the fire departments have their own helicopters.

A further application of helicopters and light aircraft is in fighting forest and bush fires, where they may be used for reconnaissance, carrying firefighting personnel, or to dump water onto the fire.

Every licenced airport has its own fire and rescue service, and the largest of all land-based fire appliances, the *crash tenders*, are to be found at big international airports. A typical example of this type of appliance is the Chubb Fire Pathfinder, weighing 37 tons and powered by a 635 brake horsepower engine. It can produce, without replenishment, 24,000 gal (109,000 l) of foam, enough to cover a Boeing 747 in a matter of seconds, applying it from its remotely controlled monitor jet on the roof.

Foam generation

Foam is generated by mixing a foaming agent with water and aerating it to produce the foam. The foaming agent may be added to the water after it

Left: An aircraft dumping fire extinguishing chemicals on a forest fire, often virtually inaccessible for land vehicles and hoses. Helicopters may also be used, for survey and rescue work.

leaves the pump, or circulated within the pump itself to mix it more effectively. The mixture is aerated at the hose nozzle by allowing air to be drawn in at the side by the vacuum created by the fast-moving stream of water and foaming agent.

Standard low-expansion foams have an expansion ratio of about 8 to 1 (the ratio of the volume of foam to the volume of the mixture before aeration), and the foaming agent is either a form of animal protein (usually hoof or horn meal), or soya protein. These foams, however, are augmented by two new types: fluoroprotein and fluorochemical synthetic detergent foams. Foams are particularly useful for dealing with liquid fires, such as oil fires, because they float on the surface of the burning liquid and so smother the flames.

High-expansion foams are based on a type of stabilized liquid detergent and can have expansion ratios of as much as 1000 to 1. They produce a very light foam which will cling to most surfaces, providing a very effective coating to extinguish the fire. In the case of a fire in a stack of tires, for example, water would put out the flames but there may be enough residual burning within the tire to start them again as the water drains off. Large amounts of water must therefore be sprayed onto the stack to prevent reignition. If high-expansion foam is used, however, it does not drain off so rapidly, and due to this fact and to the high expansion ratio of the foam, considerably less water is needed.

Extinguishers and sprinklers
The main types of portable fire extinguishers are water, foam, dry chemical and halon bromochlorodifluoromethane (BCF), which has replaced the older carbon tetrachloride type.

Water extinguishers are widely used. Previously they were operated by the interaction of a solution of sodium bicarbonate in water and a small quantity of sulfuric acid released when the extinguisher was turned upside down. Today, all extinguishers are designed to be used upright. They work by simply

Left: A sprinkler used at a fire research establishment during tests on suspended ceiling fittings. Many new homes are now built with automatic sprinklers like these installed in the ceilings.

withdrawing the safety pin and squeezing together two levers. This releases the firefighting agent, which may be continually pressurized by dry air or nitrogen (maintained-pressure type) or by carbon dioxide (CO_2) stored in a small pressure cylinder within the extinguisher shell.

Foam extinguishers use a stabilized protein or synthetic foam liquid mixed with water to be discharged as a solution to an aspirating nozzle where air is entrained to form a firefighting foam. Dry chemical types discharge a variety of special, easy-flow powders as high-velocity streams to inhibit flame chain reaction.

The carbon dioxide extinguisher consists of a high-pressure gas container filled with liquid CO_2 which is discharged via a tapered horn to form a vaporized cloud of inert gas.

Halon extinguishers generally use BCF. This is stored as a liquid and is usually pressurized to 145 psi (10 bar) to be discharged as a long-reach jet before expanding into a heavy flame-inhibiting vapor. As with CO_2 and powder extinguishers, the agent is nonconducting and can be safely used on electric equipment.

Detectors
Most of the calls dealt with by fire services are received by telephone, but increasing numbers of industrial, commercial and public buildings are being protected by automatic fire detection systems which locate the source of the fire and transmit the information to the nearest fire station. Many of these buildings also have automatic extinguishing systems that spray water, inert gas or foam directly onto the fire.

The simplest types of fire detectors use heat detectors which operate when a certain temperature is reached or the temperature rises abnormally quickly due to the heat from the fire, or *fusible links* made of a soft metal alloy that melts at these temperatures to activate the circuit. Newer systems use photoelectric cells and other devices to detect the presence of smoke, and the very latest detectors use infrared beams. A beam is transmitted across the area to be protected from transmitter to receiver. Some beam detectors use mirrors to reflect the beam to cover the area. The beam is aimed at a receiver having a grid of detectors; under stable conditions the point of contact of the beam on the detectors remains stationary, but a variation in air temperature due to a fire is enough to make the beam waver, and the detectors will react to this and trigger the alarm.

Sprinkler systems in buildings usually consist of pipes placed near the ceiling, with sprinkler heads placed at intervals along them. The heads are sealed off with fusible links, or small glass bulbs filled with a liquid or solid material that expands

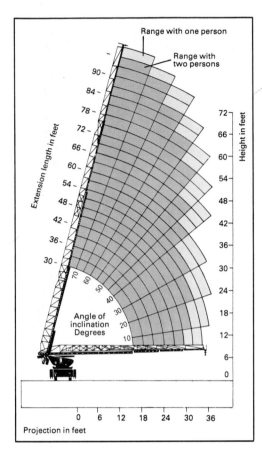

Range with one person

Range with two persons

Extension length in feet

Height in feet

90
84
78
72
66
60
54
48
42
36
30

72
66
60
54
48
42
36
30
24
18
12
6
0

Angle of inclination Degrees

70 60 50 40 30 20 10

Projection in feet

0   6   12   18   24   30   36

Above: This diagram shows the lateral and vertical extended reach of a typical rotating hydraulic ladder. The reach is greater with a single operator.

Above: Turntable (aerial) ladders being used at a paint factory fire in Maryland to extend fire hoses to reach the upper floors.

readily when the temperature rises, breaking the bulb and opening the head.

Firefighting hazards

In addition to the obvious hazards of heat, flame, smoke and falling debris, firefighters are increasingly being faced with new dangers which result from modern materials and technology. Typical of the hazards encountered even in small domestic fires are poisonous gases which are given off by plastic tiles and insulating materials, fierce and rapid flame spread from modern furnishing materials, explosive liquids and gases in aerosol cans and portable gas cylinders, and radioactive materials.

Firefighters may also be faced with any one of the hundreds of dangerous chemicals now in use, many of which are transported in bulk by road and rail. In order to face these dangers safely they need a system of communications that can bring them expert advice and information as quickly as possible.

The evolution of the present-day fire services has taken a long time, and the job will always involve a degree of personal danger for the firefighters.

See also: Flame and flashpoint; Pump.

• FACT FILE •

- Caesar Augustus formed what was probably the first municipal fire department, in Rome. Seven squads of men were led by a fire chief called a praefectus vigilum with his own chariot.

- In New Amsterdam, later to be New York, thatched roofs were outlawed in 1647, and the following year a fire prevention ordnance was passed. Money raised from fines was used to buy equipment.

- On the morning of October 8, 1871, the Chicago *Tribune* carried an insurance advertisement urging citizens to ". . . prepare now for fall and winter fires." The same day, the Great Chicago Fire broke out, which by October 10 had killed 120 people and destroyed 18,000 buildings.

Firework and flare

Fireworks probably originated as a consequence of the discovery of gunpowder in China over 2000 years ago, and were possibly used initially to frighten away devils rather than to give enjoyment to the users. The earliest European use of fireworks as an armament was by the Byzantines in the seventh century (Greek fire), but the development of fireworks for pleasure use did not begin until about AD 1500, in Italy. The practice spread throughout Europe during the sixteenth century and displays gradually became a regular feature of public entertainment on big occasions. Today fireworks are in common use for both public and private displays throughout the world.

Cases and compositions

The cases of fireworks are basically laminated paper cylinders or tubes, the thickness and shape depending on the type of firework and the composition (filling). Most cases are plain cylinder shapes, but there are variations on this, such as conical or cubic shapes, and the specialized cases for jumping crackers and Catherine wheels. The jumping cracker consists of a long thin tube folded back on itself, containing a composition designed to give a sequential series of explosions. The tube of a Catherine wheel is wound spirally around a disc of plastic, cardboard or composite material, and is consumed as the composition burns away.

The basic composition of fireworks contains compounds of potassium, carbon, and sulfur. To produce sparks, salts of lead or barium or finely powdered steel, iron, aluminum or carbon may be added to the composition. Brilliant white flame is produced by the addition of compounds of potassium, antimony, arsenic and sulfur, or powdered magnesium.

Colored flame is created by various metallic salts; strontium and lithium salts produce a red flame, green is produced by barium, yellow by sodium, and blue by copper. Colored stars and similar effects (such as those from a skyrocket) are made by small pellets of color composition which are ignited and ejected from the firework by the force composition.

Fireworks can be extremely dangerous if handled carelessly, and even more dangerous, possibly lethal, if made by amateurs. Under no circumstances should attempts be made to produce homemade fireworks, or to use fireworks in any way other than that specified by the manufacturers. Many unfortunate accidents with fireworks have ruined Fourth of July celebrations.

Displays

Portraits of personalities, depictions of buildings – sometimes as much as 600 ft (180 m) long – and moving outlines of animals or people have long been a feature of firework displays.

Among the most spectacular fireworks are shells, normally round or cylindrical paper or plastic cases projected into the air from a mortar tube sunk into the ground. These vary in size from just under 2 in. (50 mm) to 3 ft (910 mm) in diameter, and burst at a predetermined height giving star, noise and pictorial effects against the night sky. The most recent type, the daylight shell, releases slogans or product dummies on parachutes and is used for publicity and product promotion.

The preferred styles of display vary around the

Left: A fireworks display taken with a simple camera. The lens is set to a small aperture, and the film is exposed for several seconds. The red cast is given by burning chemicals.

Right: Even small domestic firework displays can make effective and breathtaking entertainment. On the Fourth of July in the U.S., on Guy Fawkes night in England, and on national days throughout the world, fireworks are set off in public areas and backyards across the country.

world, the Far East generally specializing in attaining perfect symmetry in shell-burst effects, using small short-burning stars and having a wide variety of noise effects, while in America and Europe the stars burn longer and a wider range of effects is normally produced.

Flares

The main uses of flares are for signaling (including distress signals) and for illuminating landing strips or target areas. The flares may produce a brilliant flame or colored smoke, or fire rockets into the air which release colored stars or a burning flare carried on a small parachute. The Very pistol, a widely used signaling device, fires colored stars to a height of 250 to 300 ft (76 to 91 m) from 1 or 1.2 in. (25 to 30 mm) diameter cartridges.

The main type of parachute flare is the hand-held parachute rocket. This consists of a free-flight rocket made from an aluminum alloy tube, containing the

Below: A hand-held orange flare which is used as a daytime distress signal. It burns for about 30 seconds, can be used safely even in inflatable life rafts, and is visible many miles away.

Below: A hand-held distress flare which emits a brilliant red light for about 55 seconds. The case is made of metal and the flare is operated by twisting an ignition ring.

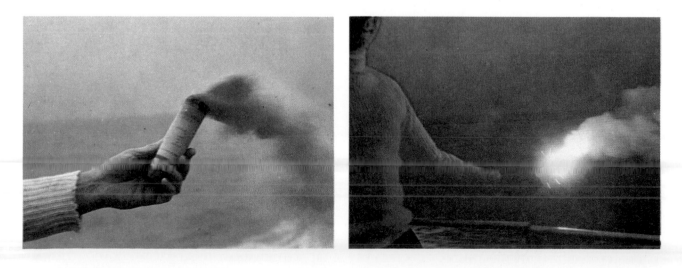

RED DISTRESS SIGNAL

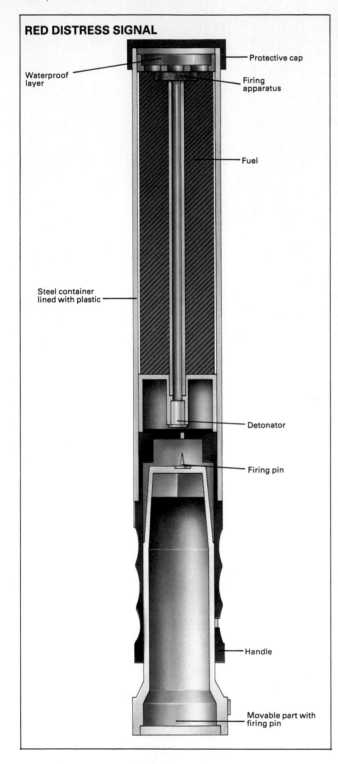

Waterproof layer

Protective cap

Firing apparatus

Fuel

Steel container lined with plastic

Detonator

Firing pin

Handle

Movable part with firing pin

Above: Distress flares are a part of the standard safety equipment of a lifeboat. The flare is lighted by the grip being turned suddenly in a clockwise direction and then the top part of the signal being pushed in forcefully. It burns for one minute with an intensity of 15,000 *candlepower,* visible from a distance of many miles.

propellant and the payload, which is a parachute-suspended flare. The rocket is fitted into a plastic launching tube and is ignited by means of a striker that fires a percussion cap. The cap in turn ignites a delay fuse and an intermediate charge known as a *quickmatch,* which ignites the propellant. The rocket motor burns for around 3.5 seconds and drives the rocket to over 1000 ft (305 m) before ejecting the parachute flare. The flare burns for over 40 seconds, and is carried on a four-string parachute which slows its rate of descent to approximately 15 ft/sec (4.6 m/sec). On a clear night such flares can be seen for a distance of about 28 miles (45 km). Illuminating parachute flares may also be dropped from helicopters or low-flying aircraft to light up large ground areas.

Handstars are used in lifeboats and by other small craft in distress, and also by mountaineers. They eject two red stars to a height of about 150 ft (46 m), with an interval of 3 to 5 seconds between them. Each star burns for about 5 seconds. Distress *handflares* produce a bright red flame and burn for about 55 seconds.

Distress flares are always colored red, and contain a mixture of magnesium, strontium nitrate, potassium perchlorate and PVC. Illuminating flares are usually white, and a range of other colors is available for signaling. A recent development is the radar-reflecting distress rocket, which carries a payload of two red stars plus about 300,000 tiny pieces of silvered nylon which will reflect radar signals and so enable rescuers to use radar to locate craft in distress.

Smoke flares

Smoke flares are used for such purposes as daylight distress signals and ground-to-air signals. Orange smoke is used for distress signals, and one type, for use by lifeboats, is designed to float and burns for two to four minutes. The smoke is nontoxic and when burning on oil-covered water it will not ignite the oil. Small hand-held smoke flares are available which burn for about 30 seconds.

An important type of smoke flare used on merchant shipping is the lifebuoy marker. One is carried on each side of the ship's bridge, attached by a line to the lifebuoys. When the lifebuoy is thrown overboard, it pulls the marker out of its mounting bracket, igniting the smoke charges. The flotation collar of the marker contains two electric lights powered by seawater-activated batteries. The smoke burns for over 15 minutes, and the lamps stay lit for over 45 minutes.

See also: Airplane; Airport; Ammunition; Battery; Carbon; Explosives; Firefighting; Flame and flashpoint; Luminescence; Parachute; Protective clothing; Radar; Rocket; Ship; Sodium; Sulfur.